Worktext 3

DIMENSIONING
A Technical Drafting Worktext

CLIFFORD J. JOENS
Drafting Instructor
Hawkeye Institute of Technology, Waterloo, Iowa

Robert J. Brady Company
A Prentice–Hall Company
Bowie, Maryland 20715

Executive Producer: Daniel S. Kamerman
Book Production Editor: Brenda Teague

Dimensioning: A Technical Drafting Worktext

Library of Congress Cataloging in Publication Data
      Joens, Clifford J 1939-
            Dimensioning.
      (His Technical Drafting Worktext; 3)
         1. Engineering drawings—Dimensioning.
I. Title.
T353.J68 vol. 3 [T357]   604'.2'4s   [604'.2'43]
ISBN 0-87618-890-0   77-13563

Prentice-Hall International, Inc., London
Prentice-Hall of Australia, Pty., Ltd., Sydney
Prentice-Hall of India Private Limited, New Delhi
Prentice-Hall of Japan, Inc., Tokyo
Prentice-Hall of Southeast Asia Pte. Ltd., Singapore
Whitehall Books, Limited, Wellington, New Zealand
Printed in the United States of America

78   79   80   81   10   9   8   7   6   5   4   3   2   1

To my wife, Janet,
and to my children, Roxane, Greg and Lindsay.

CONTENTS

● = Problem

PREFACE

Dimensioning is the third worktext of four in a Series entitled "Technical Drafting". The Series' purpose is to help students learn the basic drafting skills involved in Mechanical Drafting. The worktexts are: *Basic Drafting Skills; Multiview Drawings; Dimensioning;* and *Auxiliary Views/Descriptive Geometry.*

The author believes that application is the most important part of learning a skill and that this application should immediately follow the explanation of the skill. For this reason, the books have been published in worktext form, utilizing elements of both workbooks and textbooks. Each book in the Series is divided into modules which cover specific drafting skills. These are divided into sub-skills which are presented with illustrated explanations. Following each explanation is one or more problems in which the student learns the sub-skill by applying it. Finally, a Learning Application Problem at the end of the module provides a comprehensive application of all the sub-skills covered in the module.

Dimensioning shows the student how to dimension the features most commonly found on machine parts and how to apply the Fundamental Dimensioning rules found in ANSI Y14.5. Machined bar stock, cast and sheet metal parts are detailed from information given on design layouts. Dimensioning is primarily decimal inch and metric.

SCOPE

The Worktext Series covers basic and intermediate drafting skills. (An optional series of 10 sound filmstrips or sound slide presentations may be used to provide supplementary technical background information, if desired). The worktexts may be used in different combinations for courses of different lengths and objectives.

Course	Approximate Length	Worktext Numbers	AV Episode Numbers
Beginning Drafting	1 Semester	1 & 2	1 − 5
General Drafting	2 Quarters	1, 2, & 3	1 − 9
Intermediate Drafting	1 Semester	3 & 4	6 − 10
Descriptive Geometry	1 Quarter	4	10
Mechanical Drafting	2 Semesters or 3 Quarters	1, 2, 3 & 4	1 − 10

The Series has been class tested during four years of development in the four quarter Mechanical Drafting Program at the Hawkeye Institute of Technology in Waterloo, Iowa. During this time, 97% of the graduates have found jobs in drafting. Although this material was designed primarily for students preparing for a drafting career, it has also been used successfully in high school exploratory drafting courses and in pre-engineering courses.

ALTERNATIVE APPROACHES

Technical Drafting AV Learning Episodes is an optional series of 10 audiovisual supplements which show how the skills developed in the Technical Drafting Worktexts relate to the real needs of industrial technical drafting. The titles are: *Introduction to Drafting, Lettering, Multiview Drawing I; Multiview Drawing II; Sectional Views; Dimensioning and the Machine Shop; Dimensioning Rules; Castings; Sheet Metal;* and *Descriptive Geometry.* The Worktext Modules for which there is a corresponding AV Learning Episode are identified in the Prerequisites section of the modules.

The presentations may be made to an entire class, to a small group, or to an individual, depending on the wishes of the instructor and the AV facilities available. This technical background information may also be presented through lecture or outside readings.

Dimensioning may be used in conjunction with AV Episodes 6. *Dimensioning and the Machine Shop,* 7. *Dimensioning Rules,* 8. *Castings* and 9. *Sheet Metal.*

TO THE STUDENT

Dimensioning is divided into six modules, each of which contains several sections. These sections are explained below.

Prerequisites describe what you should be able to do before starting a module. If you can't do everything listed in the prerequisites, refer back to the designated module for review. The prerequisites of some modules include a suggestion that additional technical background would be helpful. For these modules, see your instructor for a supplementary audiovisual presentation or outside readings.

Objectives tell the specific drafting skills covered in the module. If you feel that your past experience will enable you to do everything listed without going through the module, see your instructor.

Equipment and References list all the drafting equipment and reference books you should have at hand before you begin the work in the module.

Illustrated Explanations are step-by-step diagrams and drawings, right versus wrong examples and brief written passages which are used to explain new principles and skills.

Problems are used to reinforce the skills you learn. In most cases, an introductory problem will accompany the illustration of a new principle, and it will be followed by problems of increasing complexity. If the instructions for the problem read "remove and tape down the duplicate worksheet", carefully tear out the sheet along the perforated edge. If the instructions read "tape down a drawing sheet", you are to secure your own sheet. If you need additional explanation or assistance to complete any of the problems, see your instructor. After you have completed all the problems, check with your instructor before proceeding to the Learning Application Problems. If you have not shown that you can meet all the objectives, your instructor may give you additional assistance and problems.

Learning Applications test your mastery of the skills presented in each module. When you feel you have developed enough proficiency to meet the objectives for a module, do the learning applications problems. Before handing the problems to your instructor, check them carefully against the module objectives and the drawing evaluation checklist. When you have satisfactorily met all the objectives for a module, proceed to the next.

ACKNOWLEDGMENTS

A number of illustrations and quotations in this series have been extracted from the American National Standards Institute Y14.2, Y14.3 and Y14.5 standards with permission of the publisher, the American Society of Mechanical Engineers.

The following companies have provided drawings and photographs:

Amana Refrigeration Company, Amana, Iowa

Bantam Division, Koehring Company, Waverly, Iowa

Charles Bruning Company, Division of Addressograph-Multigraph, Mount Prospect, Illinois

Collins Radio Division, Rockwell International, Cedar Rapids, Iowa

Deere and Company, Moline, Illinois

Eugene Dietzgen Company, Chicago, Illinois

Keuffel and Esser Company, Hoboken, New Jersey

A special thanks to the many students who have played such a valuable part in the development of this material over the past four years through their many suggestions.

1

BASIC DIMENSIONING

The draftsperson may receive instructions about an object to be drawn in the form of a rough sketch, a marked-up print, or a design layout. A *rough sketch* is a freehand drawing of a simple part; a *marked-up print* is a blue-line print or an enlarged, printed copy from microfilm on which required revisions are marked. A *design layout* is an un-dimensioned drawing showing a number of closely related parts to be drawn.

PURPOSE

The purpose of this module is to introduce you to the procedure involved in making a drawing from instructions which are given on a design layout, including planning the drawing or drawings, determining the best dimensions to use and applying the dimensions.

PREREQUISITES

Before you begin this module, you should be able to:

1. Use fraction, decimal-inch and millimeter scales. (Worktext 1).
2. Apply dimensions to a drawing according to ANSI Y14.5 standards for dimensioning arrangement. (Worktext 1).
3. Apply multiview drawing principles (Worktext 2).
4. Apply sectioning principles (Worktext 2).

OBJECTIVES

Upon completion of this module, you will be able to:

1. Select the combination of views, dimensions and notes which most clearly describe an object.
2. Select dimensions based on the function of the part and the manufacturing method.
3. Apply ANSI-approved dimensioning arrangement principles for fractional, decimal-inch, millimeter linear dimensions and for angular dimensions.

EQUIPMENT AND REFERENCES

Before you begin this module, you should have:

1. Drafting equipment including fractional, decimal-inch and millimeter scales.
2. ANSI Y14.5, *Dimensioning and Tolerancing*.

1-1 PLANNING THE DRAWING

In preparing to draw an object, the first step of the draftsperson is to select a drawing sheet size that will be large enough to contain the views, dimensions and notes that will be required to describe the object.

VIEWS

The number of views needed depends on the complexity of the part. Views are selected to show the **true profile** and **visible outlines** for maximum clarity. The front and side views of the object shown in **Figure 1-1** show the true profiles and visible outlines of all the features on the object, while the top view does not. (See ANSI Y14.3, paragraph 2.3 for more detail.) If a view has no dimensions and adds nothing to the clarity of the drawing, it should be omitted.

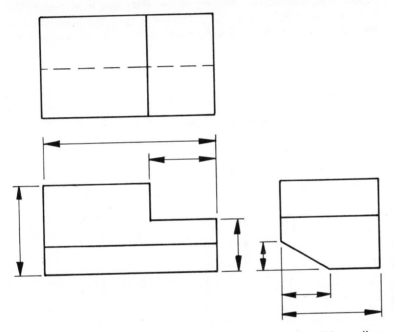

Figure 1-1. Unnecessary views. All true profiles and visible outlines are shown in the front and side views. The top view should be omitted.

DIMENSIONS

Dimensions should be shown on true profile views and should refer to visible outlines (ANSI Y14.5, paragraph 1.7g). The abbreviation "DIA" or the symbol "∅" can often eliminate the need for the circular view of a cylindrical object. A long, simple cylinder or a flat piece should be broken to save space because the dimension tells the length.

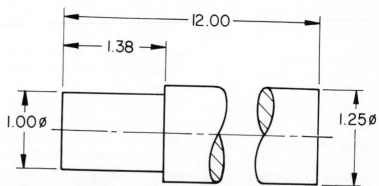

Figure 1-2. Describing with dimensions. Note that even though the circular view of the shaft is not shown and a break has been added to allow the part to be shortened to conserve space, the part is clearly and completely described.

NOTES

If the only reason for a view is to show the thickness of the material, the view can be omitted because the thickness can be specified with a note. If the description of a part consists only of stock sizes and cutoff lengths, the views can be omitted and the part described completely with a note.

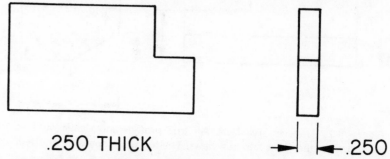

Figure 1-3. One-view drawing. The side view should be omitted with a note as shown providing the thickness dimension for the object.

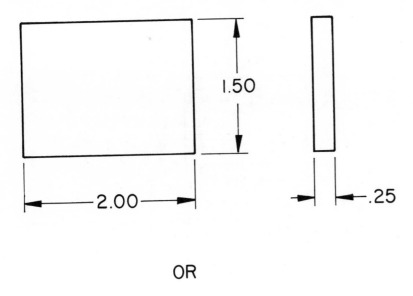

OR

.25 X 1.50 X 2.00 PLATE

Figure 1-4. Note drawing. Since the note completely describes the part, both views could be omitted. If a note is used in lieu of views, the note is sometimes placed in the middle of the drawing space of an A-size drawing sheet and sometimes given as a note, along with the part number and title, where the part is shown on an assembly drawing.

PROBLEM 1-1. Determining method of description. For the two drawings, **A** and **B**, determine the views, dimensions and notes which will most clearly describe each object and cross out the unnecessary information.

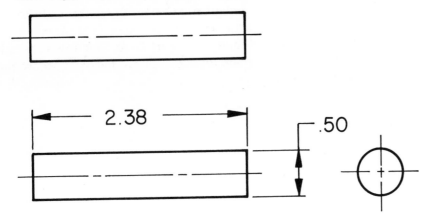

.50 ∅ X 2.38 HR STEEL

PROBLEM 1—1A

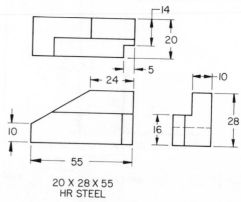

20 X 28 X 55
HR STEEL

PROBLEM 1–1B

1-2 DETERMINING THE DIMENSIONING APPROACH

After the views have been selected, the draftsperson has to decide how to dimension the part. In determining the dimensioning approach, you must first think of the function of the part, and then the manufacturing method.

FUNCTION OF THE PART

The parts must be dimensioned in a way that assures that they will fit together and work as they were intended to work. The design layout shows how the parts relate to each other, and the draftsperson has to understand these different relationships and how each part is to function with other parts. Then, he or she must dimension the parts accordingly.

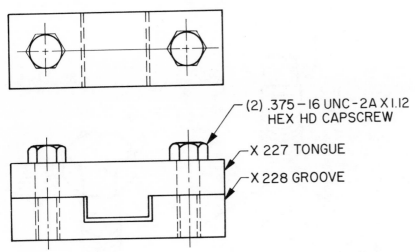

(2) .375 – 16 UNC – 2A X 1.12
HEX HD CAPSCREW

X 227 TONGUE

X 228 GROOVE

Figure 1-5. Design layout. This is the form of the information a draftsperson would receive for the detailing of parts. Parts to be detailed are identified by numbers ("X227" and "X228") and by titles ("Tongue" and "Groove"). The tongue and groove mate to form an assembly. This drawing would be scaled or measured to find the dimensions for drawing or detailing the tongue and groove on separate sheets. Since the capscrews are standard parts, they would not be detailed.

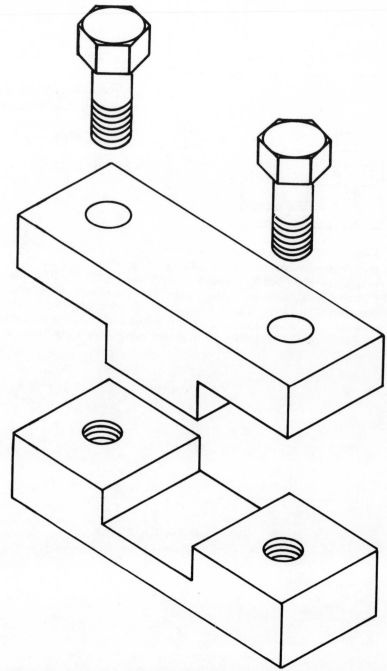

Figure 1-6. Relationship between parts. The dimensioning of the tongue and groove drawings would have to ensure that the parts would fit together and that the capscrews would be able to pass through the clearance holes in the top piece and screw into the threaded holes in the bottom piece.

MANUFACTURING METHOD

Parts should be dimensioned in a way that makes it convenient for manufacturing, provided that the method is also the best for the function of the part. In **Figure 1-7**, specifying the overall length of a part is usually best for both manufacturing and the function of the part. However, if the **1.12** and the two **1.14** dimensions were all important to the function of the part, the **3.40** overall dimension should be omitted, even though the result would cause some inconvenience to the machinist.

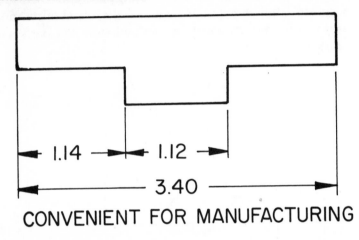

CONVENIENT FOR MANUFACTURING

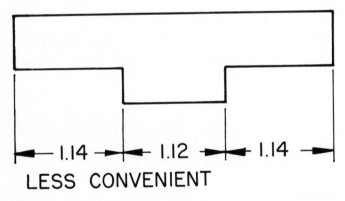

LESS CONVENIENT

Figure 1-7. Use of overall dimensions. For the part on the top, the machinist could cut the part to length without adding or subtracting dimensions. The method on the bottom is used only when the overall length is the least important dimension.

Theoretically, if mating parts such as those shown in **Figure 1—8** were specified with identical tongue and groove dimensions and a number of sets were machined, any of the tongues would fit into any of the grooves when given a light tap. In actual practice, though, some combinations would have sloppy fits and others would have to be assembled with the help of a sledge hammer because of variations in the finished products. Allowances are made for these variations in the form of clearance between mating parts and tolerances applied to the dimensions, telling the machinists and inspectors how far off each size can be before the parts will be unacceptable.

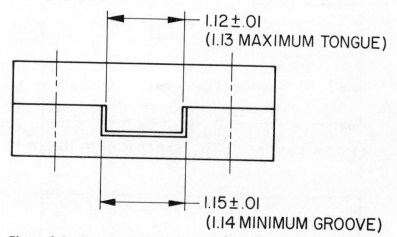

1.12±.01
(1.13 MAXIMUM TONGUE)

1.15±.01
(1.14 MINIMUM GROOVE)

Figure 1-8. Tolerances. The tolerances shown provide a minimum clearance of .01 between the mating tongue and groove. No matter what combination was put together, the groove would always be at least .01 larger than the tongue.

DETERMINATION OF TOLERANCES

Dimensions that are found by scaling a layout are usually covered by a general tolerance found in a note on the drawing. This is a tolerance that can be achieved by using normal manufacturing machines such as milling machines and lathes. If greater accuracy is required, a special tolerance must be shown by the dimension on the drawing. The draftsperson usually gets information about special tolerances from the engineer or from suppliers' catalogs for the parts that require the close-toleranced openings.

| | PLUS AND MINUS DIMENSIONS | | LIMIT DIMENSIONS | |
	SIZE	TOLERANCE	MAXIMUM	MINIMUM
EXAMPLE	1.00	±.01	1.01	.99
1	.50	±.02		
2	.749	±.001		
3	.750	±.016		
4		±	.34	.32
5		±	.375	.373
6		±	.375	.374

PROBLEM 1-2. Interpreting tolerances. The same dimension may be expressed with either a plus and minus tolerance or with maximum and minimum sizes. In the space provided, give the plus and minus or the limit dimension version for each.

1-3 MEASURING SYSTEMS

FRACTIONAL DIMENSIONS

When scaling a layout to determine a fractional size, remember that the dimensions on your part should be what the designer intended. The preferred fractional divisions are halves, quarters, eighths, and sixteenths. Thirty-seconds are used mostly where there is a small amount of clearance between parts as between a tongue and groove or between a bolt and a clearance hole. Sixty-fourths are seldom used and, if a feature on a design layout scales 49/64, you can usually assume that the designer had intended 3/4.

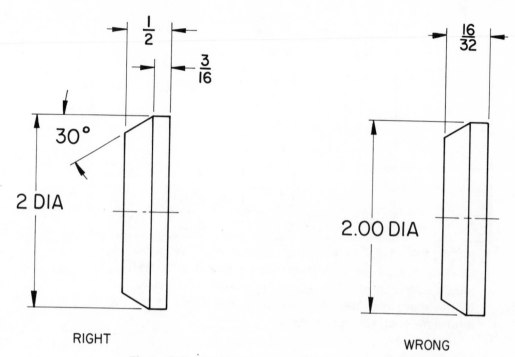

Figure 1-9. Fractional dimensions. Always reduce fractions to their lowest terms.

DECIMAL-INCH DIMENSIONS

Since the divisions on most decimal-inch scales are even hundredths, the most practical way to express a decimal-inch dimension is in even hundredths. Another advantage of even hundredth dimensions over odd hundredths is that they can be divided by two without resulting in a size expressed in thousandths. However, consideration must be given to the fact that *standard* inch sizes are based on fractional rather than decimal inches. Decimal-inch dimensions for drilled hole, thread and round stock diameters, for sheet, plate, flat and strip thicknesses, and for flat and strip widths should be converted from fractions to allow the use of standard rather than special size tools and stock. For example, a plate thickness may scale .24 with a decimal-inch scale, but since plate is not available in that thickness, .25, the conversion for 1/4 should be specified.

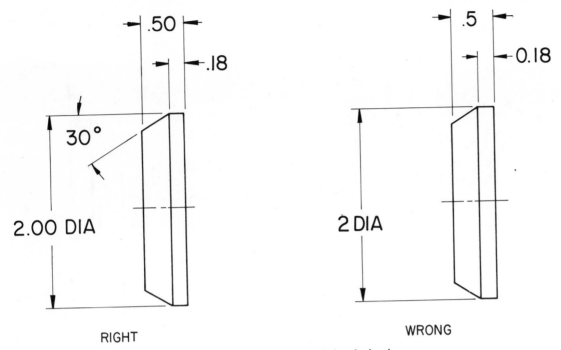

RIGHT

WRONG

Figure 1–10. Decimal-inch dimensions. Add zeros behind the decimal point of decimal-inch dimension figures but not in front of the decimal point.

MILLIMETER DIMENSIONS

The approaches to metric dimensioning range from giving the metric equivalent for standard inch sizes of everything to using metric sizes for everything. A common procedure that is used when a design is all or nearly all metric is to dimension the drawing in millimeters and to show the decimal inch equivalent of each dimension in a chart on the left side of the drawing. This conversion chart may be used when metric machines and measuring equipment are not available for the manufacture of the part.

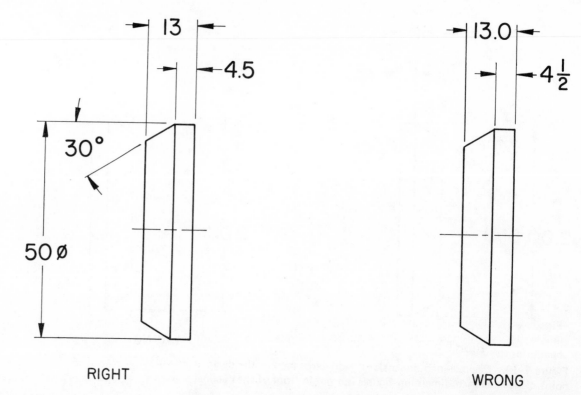

Figure 1-11. Millimeter dimensions. Add a zero in front of the decimal point of a millimeter dimension figure but not behind the decimal point.

1-4 DIMENSIONING ARRANGEMENT

LINEAR DIMENSIONS

When space allows, the dimension figure and the arrowheads should be placed between the extension lines. When this arrangement would cause cramping, one of the limited space arrangements should be used. Arrowheads must always be at least .12 long and dimension figures should *never* touch another line on the drawing.

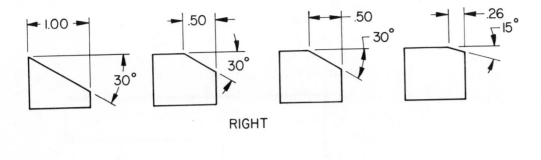

RIGHT

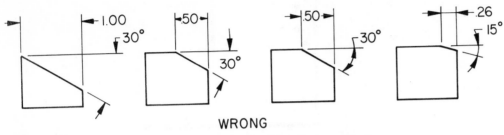

WRONG

Figure 1-12. Dimensioning arrangement. In the "right" examples, the linear and angular dimensions are arranged so that neither the arrowheads nor the dimension figures are crowded. In the "wrong" examples, crowding occurs and arrows are misplaced.

ANGULAR DIMENSIONS

The arrangement for angular dimensions is similar to that of linear dimensions. The dimension line arc must be swung from the intersection of the sides of the angle.

PROBLEM 1-3. Dimensioning arrangement. Remove and tape down the duplicate worksheets at the back of the book. Dimension the objects completely, using the indicated system of measurement.

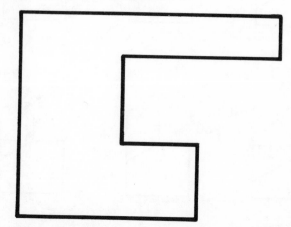

PROBLEM 1—3A Fractional inch. Specify 1/4 thickness.

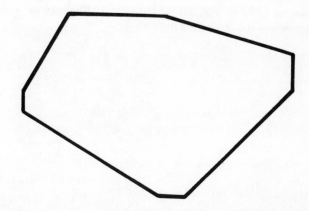

PROBLEM 1—3B Decimal inch. Specify .250 thickness.

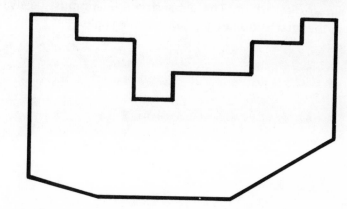

PROBLEM 1—3C Millimeters. Specify 6.00 thickness.

1-5 LEARNING APPLICATION

Your work on Problem 1-4 will show how well you have met the objectives of Module 1.

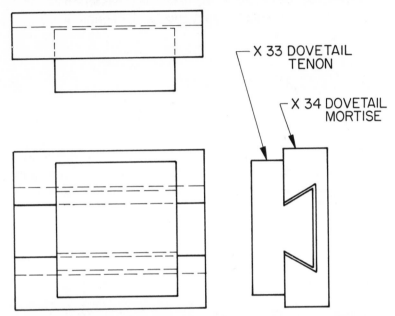

X 33 DOVETAIL TENON

X 34 DOVETAIL MORTISE

PROBLEM 1-4. Basic dimensioning. On separate drawing sheets, draw the parts shown in the given design layout according to the scaled sizes. Use the combination of views, dimensions and notes that best describe the parts. Use fractional-inch, decimal-inch or millimeter dimensions.

DRAWING CHECKLIST

1. Did you select the views that show the true profile and visible outline for maximum clarity?
2. Did you omit views with no dimensions and those which could be replaced by a note?
3. Did you dimension the parts so they would fit together and work as they were intended to?
4. Did you dimension the parts for ease in the type of manufacturing involved?
5. Did you include a general tolerance?

6. Did you use proper fractional, decimal-inch or metric dimensioning procedures?
7. Did you observe the standards for spacing of dimension lines and for extension lines?
8. Did you arrange the dimension figures and arrowheads to avoid cramping?
9. Did you use an arc swung from the intersection of the sides of the angle for your angular dimension lines?

ANSWERS TO PROBLEMS

These partial solutions contain some of the dimensions and a general approach which may be used. In some cases, other approaches may also satisfy ANSI dimensioning requirements.

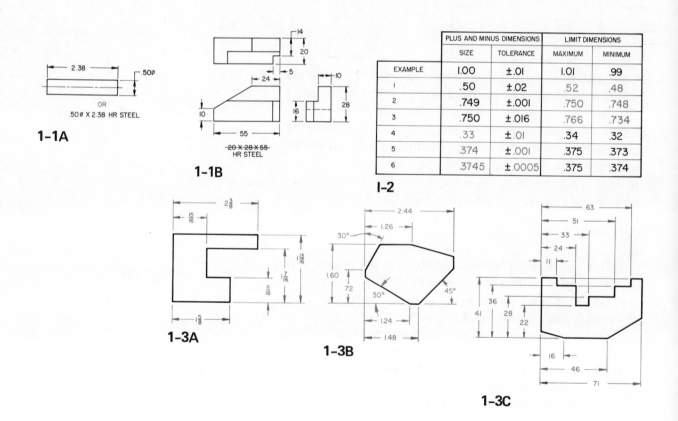

| EXAMPLE | PLUS AND MINUS DIMENSIONS | | LIMIT DIMENSIONS | |
	SIZE	TOLERANCE	MAXIMUM	MINIMUM
EXAMPLE	1.00	±.01	1.01	.99
1	.50	±.02	.52	.48
2	.749	±.001	.750	.748
3	.750	±.016	.766	.734
4	.33	±.01	.34	.32
5	.374	±.001	.375	.373
6	.3745	±.0005	.375	.374

1-1A

1-1B

I-2

1-3A

1-3B

1-3C

2

HOLES

Some of the most common features that a draftsperson will find on drawings of mechanical parts involve holes. In dimensioning holes, the draftsperson must first of all bear in mind how the part is to be used and how specific holes relate to each other and to other features on the drawing. Second, how the part will be made must be considered. The dimensions for the hole should allow the part to be produced without unnecessary difficulty and to function properly in use.

PURPOSE

The purpose of this module is to show you industry-approved methods of dimensioning the size and locations of holes and explain the concepts of *American National Standard Y14.5.*

PREREQUISITES

Before you begin this module, you should be able to:

1. Use decimal-inch and millimeter scales (Worktext 1).
2. Apply multiview drawing principles (Worktext 2).
3. Dimension simple objects with ANSI approved dimensioning arrangements (Module 1).

OBJECTIVES

Upon completion of this module, you will be able to:

1. Dimension the location of holes using circular centerlines, chain dimensioning and baseline dimensioning.
2. Dimension the size of drilled, punched, countersunk, counterbored, spotfaced, threaded, bored and cored holes.
3. Dimension holes on mating parts.

EQUIPMENT AND REFERENCES

Before you begin this module, you should have:

1. Drafting equipment including drafting machine, decimal-inch scale, metric scale, circle template, compass and lead holder.
2. American National Standard Y14.5, *Dimensioning and Tolerancing*.
3. *Machinery's Handbook* (optional reference).

2-1 HOLE LOCATION DIMENSIONING

LOCATING DIMENSIONING METHODS

A hole is always located by its centerline (**Figure 2-1**), preferably in the circular view (the view in which it appears as a circle). Refer to ANSI Y14.5, paragraph 1.7g.

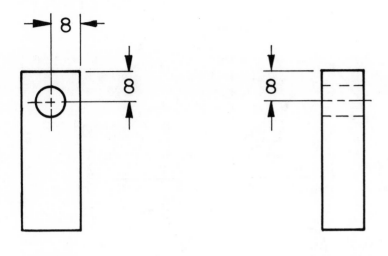

RIGHT   WRONG

Figure 2-1. Hole location dimensions. In the "wrong" example the hole is located in the rectangular view.

Circular Centerline Dimensioning. A circle of centers or bolt circle is a circular centerline for holes which are equidistant from a center. See **Figure 2-2.**

Figure 2-2. Circular centerline dimensioning. If there are equal angles between the holes, "EQL SP" (equally spaced) is indicated in the note providing the hole size (A) and (B) rather than giving angular dimensions.

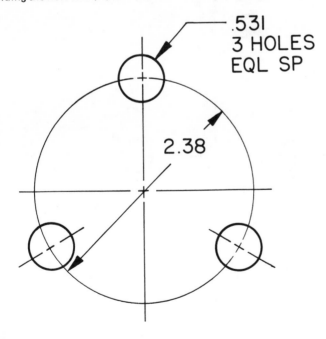

.531
3 HOLES
EQL SP

2.38

(A)

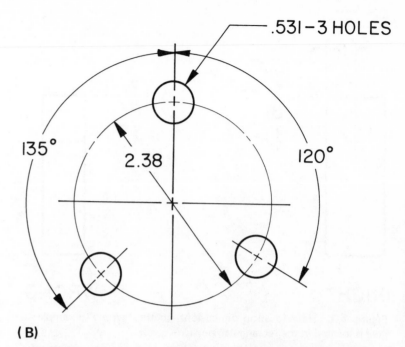

.531–3 HOLES

135°

2.38

120°

(B)

PROBLEM 2-1. Circular centerline dimensioning. Remove and tape down the duplicate worksheet at the back of the book. Dimension the locations of the holes in decimal inch according to the examples shown in **Figure 2-2.**

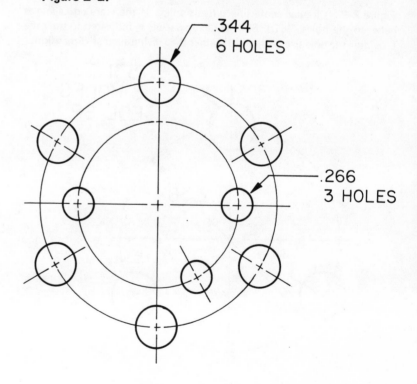

.344
6 HOLES

.266
3 HOLES

Chain Dimensioning. This method gives the locations of holes from center to center (**Figure 2-3**). If the holes are located from opposite ends, the parts are harder to produce and assemble.

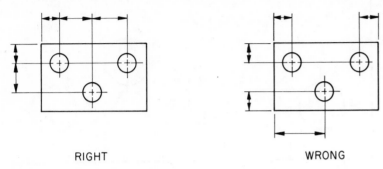

RIGHT   WRONG

Figure 2-3. Chain dimensioning. In the "right" example the holes are "tied together" by dimensions between the holes. In the "wrong" example, the holes are located from all sides and not from each other.

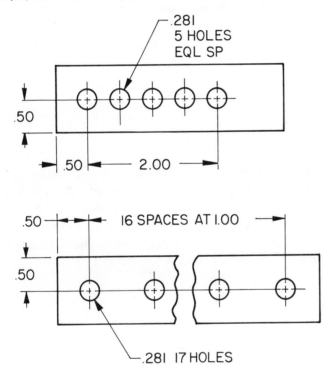

Figure 2-4. Equally spaced holes. Using one of these methods for locating equally spaced holes can save time and still clearly convey the intent of the drawing.

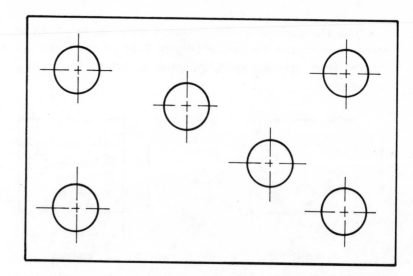

PROBLEM 2-2. Chain dimensioning. Remove and tape down the duplicate worksheet at the back of the book. Dimension the location of the holes according to the example shown in **Figure 2-3.** Scale the drawing with a fractional scale and give the dimensions to the nearest sixteenth.

Baseline Dimensioning. Baseline dimensioning or rectangular coordinate dimensioning of holes gives the locations from vertical and horizontal baselines (**Figures 2-5, 2-6** and **2-7**). These baselines must extend from smooth, accurate surfaces because dimensions can never be more accurate than the surface from which they are measured.

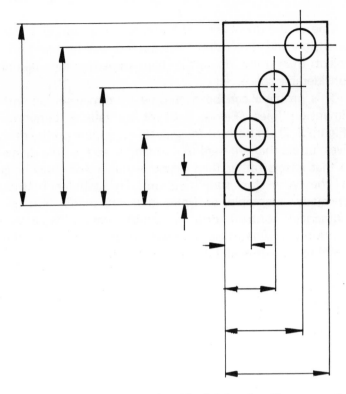

Figure 2-5. Baseline dimensioning. The left-hand and bottom surfaces are the "baseline" or "datum" surfaces.

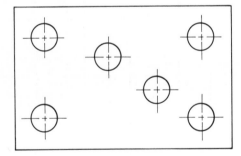

PROBLEM 2-3. Baseline dimensioning. Remove and tape down the duplicate worksheet at the back of the book. Dimension the locations of the holes according to the example shown in **Figure 2-5.** Scale the drawing with a decimal-inch scale and give the dimensions to the nearest even hundredth.

One of the drawbacks of baseline dimensioning is that the extension and dimension lines extend out further from the object than they do with chain dimensioning, taking up considerable space.

This problem can be overcome by dimensioning without dimension lines (**Figure 2-6**) or by tabular dimensioning (**Figure 2-7**). In both of these systems, a table on the drawing gives letters that are used to correspond to the hole diameters so that letters rather than diameter dimensions may be given in the view for simplification. The tabular method is especially adaptable to holes that are made with an NC (numerical control) drill or punch press. The X and Y location coordinates can be used directly to program the NC machines to locate all holes automatically.

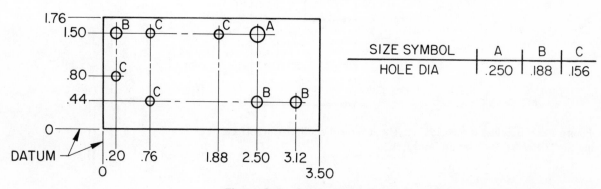

SIZE SYMBOL	A	B	C
HOLE DIA	.250	.188	.156

Figure 2-6. Dimensioning without dimension lines. In this form of baseline dimensioning, the baselines are marked "O". This method saves time and space on the drawing.

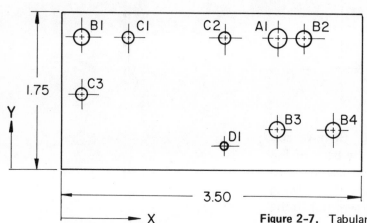

REQD		I	4	3	I
HOLE DIA		.250	.188	.156	.125
POSITION		HOLE SYMBOL			
X →	Y ↑	A	B	C	D
2.50	1.50	AI			
.19	1.50		BI		
2.81	1.50		B2		
2.50	.44		B3		
3.00	.44		B4		
.75	1.50			CI	
1.88	1.50			C2	
.19	.81			C3	
1.88	.25				DI

Figure 2-7. Tabular dimensioning. This method is especially useful on drawings of parts to be made on numerical control machines.

REQD		2	4	3
HOLE DIA		20.00	16.00	10.00
POSITION		HOLE SYMBOL		
X →	Y ↑	A	B	C
58	76	A1		
104	58	A2		
16	86		B1	
116	86		B2	
116	16		B3	
16	16		B4	
30	58			C1
58	40			C2
74	18			C3

PROBLEM 2-4. Dimensioning without dimension lines. Tape down a drawing sheet. Draw a 102mm X 132mm (height X width) plate and add the holes described in tabular form above. Dimension the size and location of the holes without dimension lines, according to the example shown in **Figure 2-6.**

2-2 HOLE SIZE DIMENSIONING

DRILLED AND PUNCHED HOLES

Holes may be made by drilling, punching, reaming, boring and other methods. If the material in which the hole is to be made is over .12 thick and the hole is under 1.00 in diameter, the most economical and commonly used process for making the hole is drilling. For stamped sheet-metal parts, the most commonly used process for making holes is punching.

Standard Drilled Hole Sizes. For holes in which drilling is a likely choice for the production method, hole size dimensions should conform to standard drill sizes (**Figure 2-8**). Drills in the inch system are available in the fraction, number and letter sizes shown in Appendix I at the back of this book. Appendix II provides standard metric drill sizes.

.047	.050
.201	.200
.234	.230
STANDARD INCH DRILL SIZES	NOT STANDARD

6.00	6.10
12.50	12.80
STANDARD METRIC DRILL SIZES	NOT STANDARD

Figure 2-8. Standard drilled hole sizes. The sizes marked "not standard" cannot be made within a standard drill and should be avoided. Use sizes from standard drill tables if possible.

Drilled and Punched Hole Notes. The size of a drilled or punched hole should preferably be dimensioned by note in the circular view (**Figure 2-9**). If there is no circular view, the hole is dimensioned in the rectangular view and the abbreviation "DIA" or the symbol "∅" is given after the size. Drawings should give the size and tolerance of holes to tell the production or machine shop the required end results without dictating the process to be used. Although the tolerances may be applied directly to the applicable dimension, they are generally given in a note along with the other general tolerances. Hole notes also include the depth of blind holes (holes that do not go completely through the material) and the quantity of holes if there is more than one. Hole depths apply to the full diameter and not to the point of the drill.

Figure 2-9. Drilled, punched and reamed hole specifications. In the "right" example, the sizes of the holes are given without dictating the method of manufacture. Any process may be used, so long as the tolerances are met. In the "wrong" example, the processes have been specified.

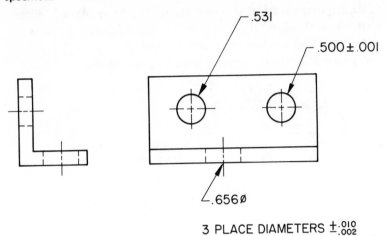

3 PLACE DIAMETERS $\pm^{.010}_{.002}$

RIGHT

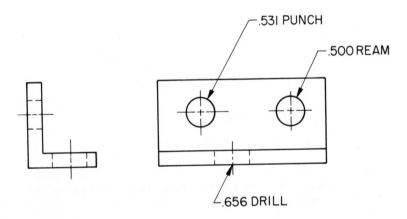

WRONG

Leaders. A leader line used in dimensioning starts with a short horizontal shoulder placed in front of the first or in back of the last letter or digit of the hole-size note. It extends from the shoulder toward the center of the hole, terminating with an arrowhead at the outside of the hole. Avoid leaders that cross leaders that extend in a horizontal or vertical direction, and leaders that are parallel to nearby dimension lines, extension lines, section lines and centerlines.

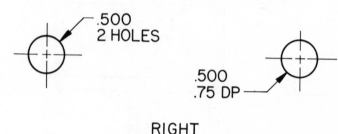

RIGHT

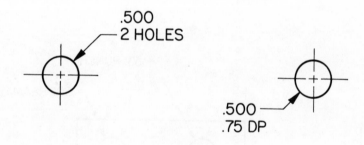

WRONG

Figure 2–10. Leader shoulders. In the "right" examples, the shoulder extends from the front of the top line in one case and the back of the bottom line in the other. In the "wrong" examples, the shoulder extends from the front of the bottom line in one case and the back of the top line in the other.

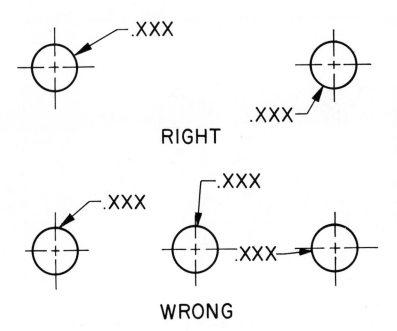

RIGHT

WRONG

Figure 2—11. Leader angles. The leader should preferably be angled between 30° and 60°.

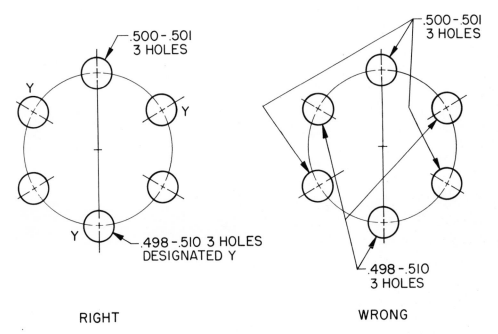

RIGHT

WRONG

Figure 2-12. Leader systems. Confusion can be minimized by including the quantity of holes in the note and by labeling holes which are nearly the same size.

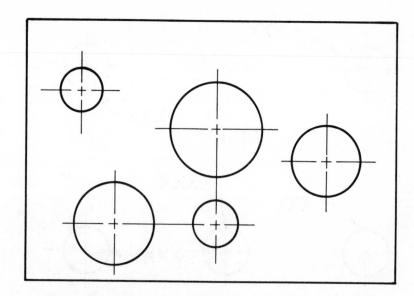

PROBLEM 2–5. Specifying inch drilled hole sizes. Remove and tape down the duplicate worksheet at the back of the book. Scale the diameters of the holes and find the nearest standard inch drill size in Appendix I in the back of the book. Dimension the sizes of the holes in thousandths and the locations in even hundredths of an inch (decimals).

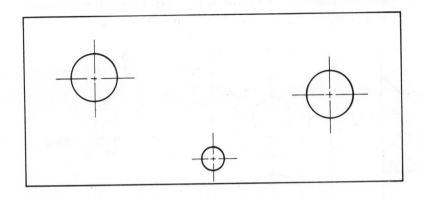

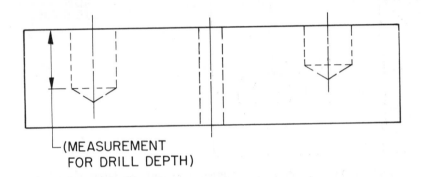

└─(MEASUREMENT
FOR DRILL DEPTH)

PROBLEM 2-6. Specifying drilled hole sizes (metric). Remove and tape down the duplicate worksheet at the back of the book. Scale the diameters of the holes and find the nearest standard millimeter drill size in Appendix II at the back of the book. Dimension the sizes of the holes in two-place millimeters and the location and depths of the holes in whole millimeters.

COUNTERSUNK, SPOTFACED AND COUNTERBORED HOLES

These types of holes are exceptions to the rule about describing the end result rather than dictating the process. The terms **CSK, S FACE** and **C BORE** are used because they describe the shape of the holes as well as the processes. Each of these features is used in conjunction with a hole through which a bolt or capscrew is to pass. *It is essential that the drilled hole be larger than the bolt or capscrew so that the fastener will fit through.* Although practices vary from company to company and between large and small holes, clearance holes are most often 1/32 (.031) or 0.8 millimeters larger than the thread size of the bolt or capscrew.

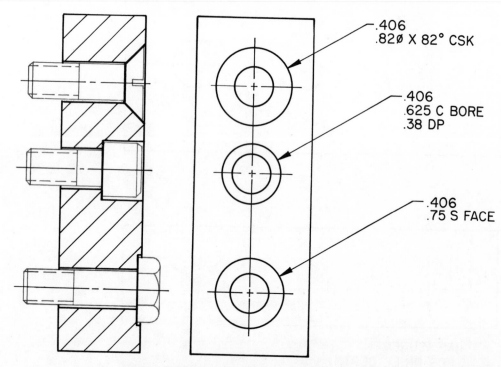

.406
.820 X 82° CSK

.406
.625 C BORE
.38 DP

.406
.75 S FACE

Figure 2-13. Countersunk, counterbored and spotfaced holes. The countersunk hole allows flush mounting of a flat-head screw; the counterbored hole allows flush mounting of a hex socket head screw; and the spotfaced hole provides a smooth, square surface for the underside of a screw head. Each of the screws has a thread size of .375 (3/8), so the clearance holes are made .406 (13/32) or 1/32 larger than the thread size.

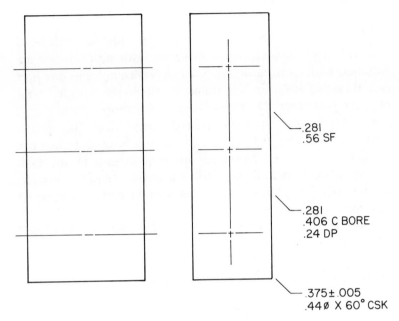

.281
.56 SF

.281
.406 C BORE
.24 DP

.375±.005
.440 X 60° CSK

PROBLEM 2- 7. Drawing CSK, C BORE and SF holes. Remove and tape down the duplicate worksheet at the back of the book. Draw the circular view and the hidden side view of each hole according to the given dimensions.

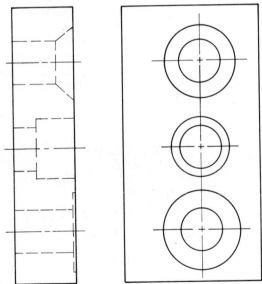

PROBLEM 2–8. Specifying CSK, C BORE and SF holes. Remove and tape down the duplicate worksheet at the back of the book. Scale the drawing and dimension the size and location of the holes according to the examples shown in **Figure 2–13.**

THREADED HOLES

While some mating parts are fastened together with bolts and nuts and have clearance holes in both parts, many are fastened with capscrews and have clearance holes in one part and threaded holes in the other. Producing a threaded hole requires two steps: (1) a drill is used to produce a hole that is somewhat smaller than the desired thread size. This drill is referred to as the "tap drill". (2) A tap is used to cut the threads in the hole. Notes for internal threads in the inch system should include tap drill diameter, thread diameter, threads per inch, thread type, class of fit and indication of internal threads.

The tap drill and threads per inch for each standard inch thread size are found in Appendix I.

Figure 2-14. Specifying inch internal threads.

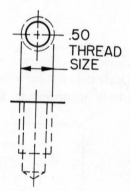

.50 THREAD SIZE

(A) Find the thread size by scaling the dashed lines in either the circular or rectangular view.

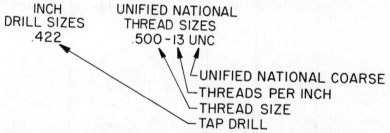

INCH DRILL SIZES
.422

UNIFIED NATIONAL THREAD SIZES
.500-13 UNC

└UNIFIED NATIONAL COARSE
└THREADS PER INCH
└THREAD SIZE
└TAP DRILL

(B) Look up the thread size in Appendix I at the back of the book, selecting the size that is closest to one of the given Unified National Thread Sizes. Although both UNC (Unified National Coarse) and UNF (Unified National Fine) are shown, UNC is the most commonly used and you should assume the thread is UNC unless you are given other instructions.

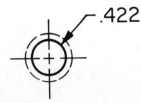

(C) Find the tap drill that corresponds with the thread size and specify it on your drawing. Since the tap drill is not drawn exactly to scale, always look up the tap drill in a table rather than using the scaled size.

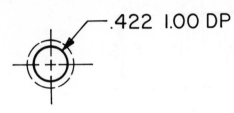

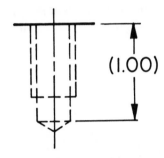

(D) Scale the drawing to determine the depth of the tap drill and specify it after the size specification.

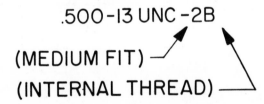

(E) Specify the thread as it is shown in Appendix I. In addition, include the class of fit and the designation for internal thread. "2B" indicates a medium fit internal thread. The information from the thread size through the indication of internal thread must be placed on one line.

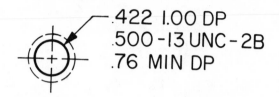

.422 1.00 DP
.500-13 UNC-2B
.76 MIN DP

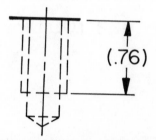

(.76)

(F) Scale the drawings to determine the depth of threads and specify as MIN DP. The thread depth is always less than the drill depth.

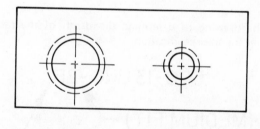

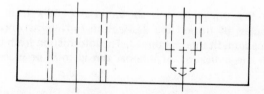

PROBLEM 2-9. Specifying inch threads. Remove and tape down the duplicate work sheet at the back of the book. Dimension completely, specifying the threaded holes according to the scaled sizes and the steps shown in **Figure 2-14.**

Notes for metric and internal threads should include tap drill diameter, thread diameter, thread pitch, class of fit and indication of internal threads. The tap drill and pitch for the most commonly used metric thread sizes are found in Appendix II. For more detailed information on inch and metric threads, consult the *Machinery's Handbook* or some other general reference source.

Figure 2-15. Specifying metric threads. The procedure for specifying metric threads is, for the most part, the same as specifying inch threads.

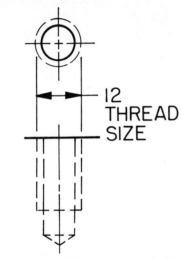

12
THREAD
SIZE

(A) Determine the thread size the same as inch threads.

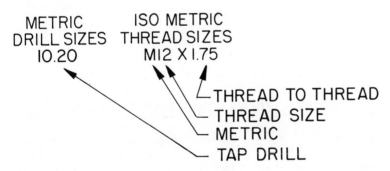

METRIC
DRILL SIZES
10.20

ISO METRIC
THREAD SIZES
M12 X 1.75

THREAD TO THREAD
THREAD SIZE
METRIC
TAP DRILL

(B) Look up the thread size in Appendix II at the back of the book. This step is the same as inch threads except there are no coarse and fine designations.

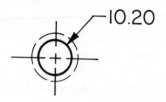

(C) Find the tap drill the same as the inch sizes and specify it on the drawing.

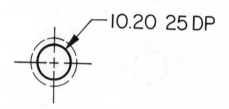

(D) Specify the drill depth the same as the inch system.

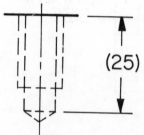

(E) Specify the thread as it is shown in Appendix II. Instead of threads per inch as found in the inch threads, the pitch is the distance from one thread to the next thread in millimeters. In addition, include the class of fit and the designation for internal thread. "6H" indicates a medium fit internal thread.

10.20 25 DP
M12 X 1.75−6H
19 MIN DP

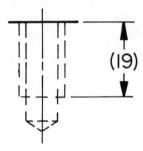

(19)

(F) Specify the thread depth as in the inch system.

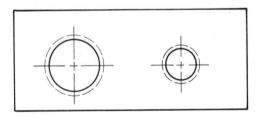

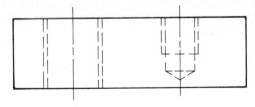

PROBLEM 2-10. Specifying metric threads. Remove and tape down the duplicate worksheet at the back of the book. Dimension the object completely, specifying the threaded holes according to the scaled sizes and the steps shown in **Figure 2-15.**

BORED AND CORED HOLES

Machined holes above one inch are most often made by a process called boring. Non-machined holes in cast iron are made with cores in the casting process (**Figures 2–16** and **2–17**). The size dimensions in these non-standard holes may be given in either the circular or rectangular view, whichever shows the shape most clearly. Large concentric holes are most clearly shown in a *sectioned* rectangular view. If there is no circular view, the dimension should be followed by the abbreviation "DIA" or symbol "∅".

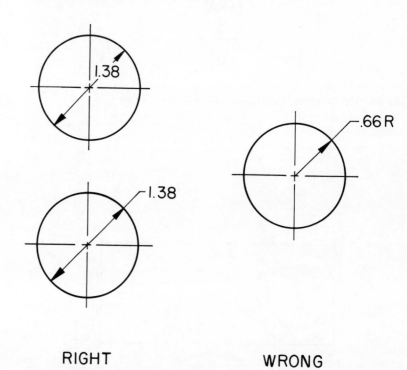

RIGHT   WRONG

Figure 2–16. Circular view dimensioning of large holes. In the "right" examples, the holes are dimensioned by diameter, with the dimension figure positioned so that crowding is avoided. In the "wrong" example, the diameter is dimensioned by its radius.

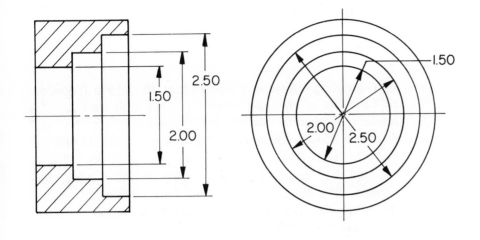

CLEAR   UNCLEAR

Figure 2-17. Rectangular view dimensioning of concentric holes. The holes are more clearly shown in the sectioned rectangular view. The circular view may be omitted if the "∅" symbol is placed behind the dimensions in the rectangular view.

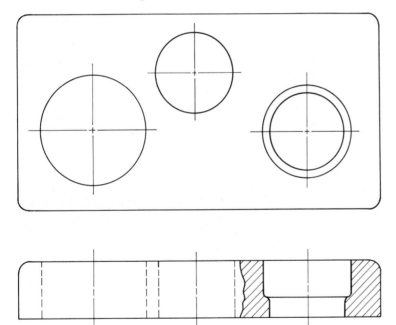

PROBLEM 2-11. Dimensioning non-standard holes. Remove and tape down the duplicate worksheet at the back of the book. Dimension the holes according to the scaled sizes and the examples shown in **Figure 2-16** and **Figure 2-17.**

In deciding how to dimension a part, the intended purpose or function of the part must be considered first and the method of manufacture second. It is essential that bolt and capscrew holes be dimensioned so that the mating holes will line up and allow the bolts to pass through. This is accomplished partly by the practice of "like parts, like dimensions", dimensioning mating hole patterns in exactly the same way with identical location dimensions on both parts wherever possible.

Figure 2–18. Mating holes.

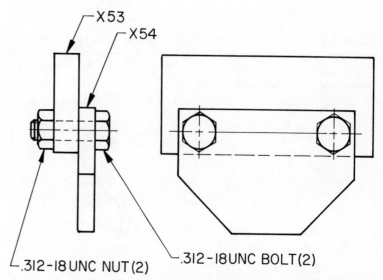

(A) Mating parts **X53** and **X54** are fastened with **.312-18** bolts and nuts.

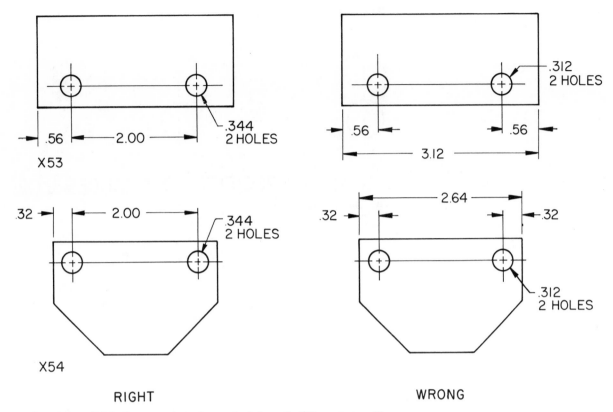

RIGHT   WRONG

(B) In the "right" example, the principle of "like parts, like dimensions" is applied between the holes and clearance is provided for the bolts. In the "wrong" example, the location dimensions do not match and the hole sizes provide no clearance.

CLEARANCES AND TOLERANCES

If the holes in mating parts are drilled in exactly the locations that are specified on the drawings, the parts will always fit together. However, it is impossible for the shop to make perfect parts, and unless some provision is made for slight inaccuracies, the parts will not fit together. These provisions for inaccuracy are called clearances and tolerances. If the hole is made 1/32 larger than the thread of the bolt, there is clearance between them that allows the location of the holes to vary a little without preventing the bolts from going through the mating holes. This allowable variation in the location of the holes is called a tolerance. While special tolerances are applied directly to some dimensions on some drawings, the majority of the dimensions will be covered by general tolerances given in a note on the drawing.

Figure 2–19. Interpretation of standard tolerances. Here are three ways of applying a tolerance to a dimension.

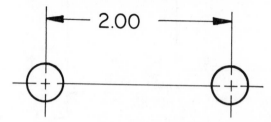

UNSPECIFIED TOLERANCES±.02

(A) Use of general note.

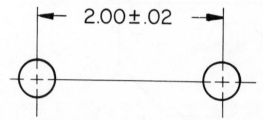

(B) Applying tolerance directly to the dimension.

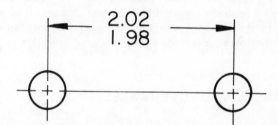

(C) Expressing the toleranced dimension as maximum and minimum sizes.

ACCUMULATION OF TOLERANCES

The disadvantage of chain dimensions is that the tolerances can build up, or accumulate. If all the dimensions on one part are the smallest sizes that are permissible according to the tolerances, and the dimensions on the mating part are the largest sizes that are permissible, the parts could not be assembled since alignment would become progressively worse. The major advantage of baseline dimensioning is that there is a minimum of tolerance accumulation.

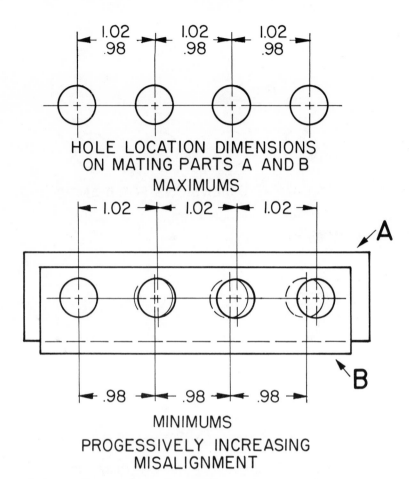

HOLE LOCATION DIMENSIONS
ON MATING PARTS A AND B
MAXIMUMS

MINIMUMS

PROGESSIVELY INCREASING
MISALIGNMENT

Figure 2-20. Tolerance accumulation with chain dimensions. The misalignment increases with each hole.

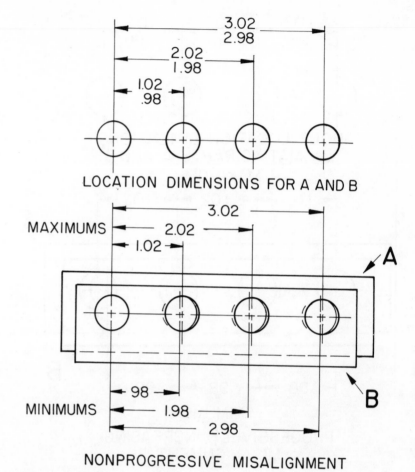

LOCATION DIMENSIONS FOR A AND B

MAXIMUMS

MINIMUMS

NONPROGRESSIVE MISALIGNMENT

Figure 2-21. Tolerance accumulation with baseline dimensioning. The misalignment remains constant, regardless of the number of holes.

ITEM	PART NUMBER	DESCRIPTION
I	X252	PLATE
2	X253	COVER
3	STANDARD	(4) .250-20 UNC HEX HD MACH SCR

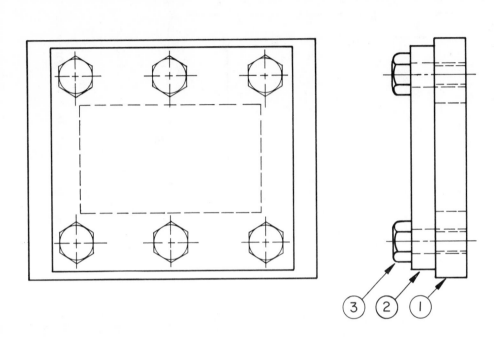

PROBLEM 2-12. Mating parts dimensioning. On two separate drawing sheets, draw and completely dimension parts **X252** and **X253**, according to sizes scaled in the given drawing. Use Appendix I for the threaded hole specifications.

2-4 LEARNING APPLICATION

Problem 2-13 will show how well you have met the module objectives.

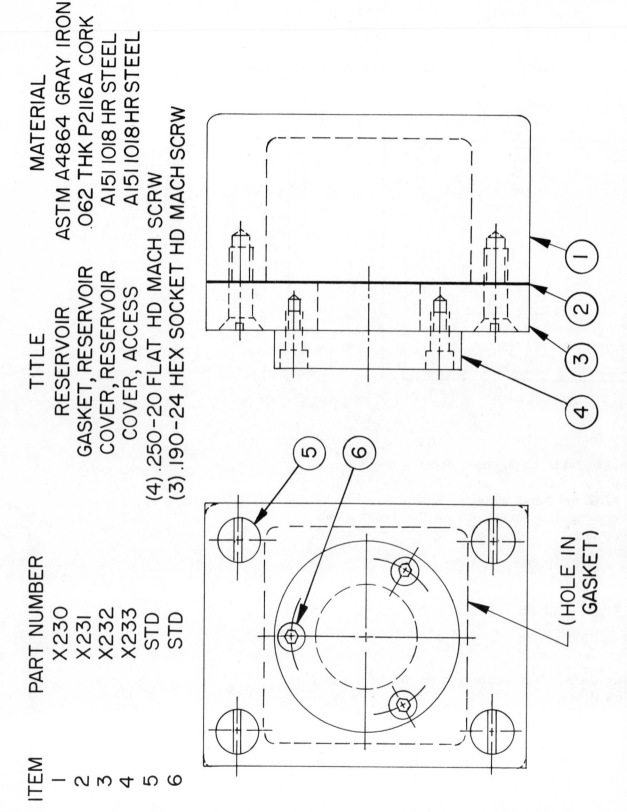

ITEM	PART NUMBER	TITLE	MATERIAL
1	X230	RESERVOIR	ASTM A4864 GRAY IRON
2	X231	GASKET, RESERVOIR	.062 THK P2116A CORK
3	X232	COVER, RESERVOIR	AI51 1018 HR STEEL
4	X233	COVER, ACCESS	AI51 1018 HR STEEL
5	STD	(4) .250-20 FLAT HD MACH SCRW	
6	STD	(3) .190-24 HEX SOCKET HD MACH SCRW	

(HOLE IN GASKET)

PROBLEM 2-13. Dimensioning holes. On three separate drawing sheets, draw and completely dimension parts **X231**, **X232** and **X233**, according to sizes scaled in the given drawing. Use Appendix I for threaded hole specifications.

DRAWING EVALUATION CHECKLIST

1. Do all mating hole patterns have the same location dimensions ("Like parts, like dimensions")?
2. Are the clearance holes 1/32 (.031) larger than the thread size of the fastener that goes through them?
3. Are the drilled holes standard?
4. Are the countersinks dimensioned in the standard way?
5. Are the counterbores dimensioned properly?
6. Is the correct tap drill and thread specification used for the threaded holes?

ANSWERS TO PROBLEMS

These partial solutions contain some of the dimensions and a general approach which may be used. In some cases, other approaches may also comply with the ANSI dimensioning requirements.

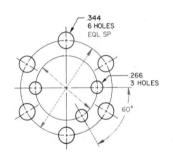

2-1

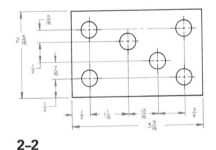

2-2

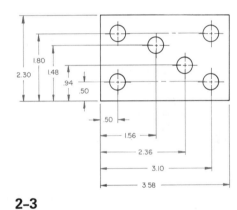

2-3

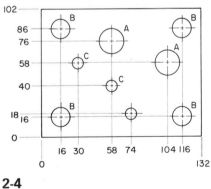

2-4

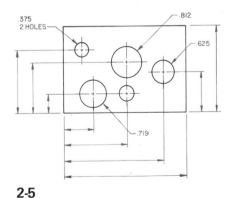

2-5

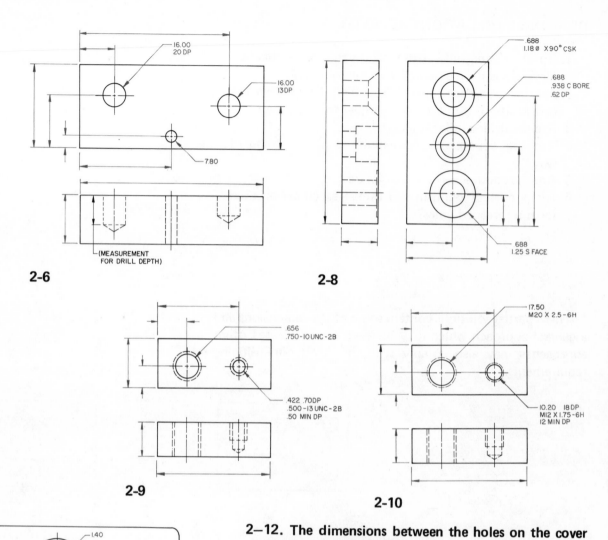

2-6

2-8

2-9

2-10

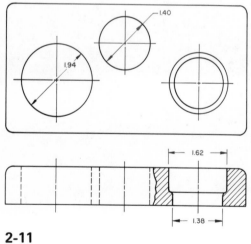

2-11

2—12. The dimensions between the holes on the cover must be exactly the same as the dimensions between the drilled and tapped holes in the plate. The hole size note in the cover should be:

.281 6 HOLES

The hole size note in the plate should be:

.201 (7)
.250—20 UNC—2B
6 HOLES

3

COMMON FEATURES

Although a group of machine parts may not resemble each other at all in size and shape, they usually have a number of features which are the same. There are many ways in which these features could be described, but the drawings will be more easily understood if similar features are dimensioned in the same way on every drawing. The ANSI Standard Y14.5 for *Dimensioning and Tolerancing* shows methods of dimensioning the most commonly found features that have been agreed upon by experts in industry, education and government. These methods have been adopted by the majority of manufacturing companies across the country and have been placed in company drafting manuals which serve as guides to practicing draftspersons and checkers.

PURPOSE

The purpose of this module is to show you the ANSI-recommended ways of dimensioning some of the most common features found on drawings.

PREREQUISITES

Before you begin this module, you should be able to:

1. Apply basic drafting skills (Worktext 1).
2. Apply multiview drawing principles (Worktext 2).
3. Apply ANSI Y14.5 standards in dimensioning lengths, angles and holes (Modules 1 and 2).

OBJECTIVES

Upon completion of this module, you will be able to:

1. Dimension the size and location of radii.
2. Dimension rounded-end and rounded-corner parts.
3. Dimension internal and external chamfers.
4. Dimension external threads.
5. Dimension round- and flat-bottom grooves.
6. Dimension straight, diagonal and diamond knurls.

EQUIPMENT AND REFERENCES

Before you begin this module, you should have:

1. All normal drafting equipment including decimal-inch and millimeter scales.
2. ANSI Y14.5, *Dimensioning and Tolerancing.*

3-1 RADII

SIZE OF RADII

An arc is dimensioned by giving its radius. For large radii, the dimension figure should be placed between the radius center and the arrowhead. For small radii, the arrangement used must provide sufficient space for full-size dimension figures and arrowheads. See **Figure 3-1.**

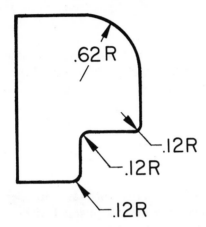

Figure 3-1. Size of radii. While there is room for a .62 R dimension between the arrowhead and arc center, a .12 R dimension must be placed beyond the arrowhead or beyond the arc center.

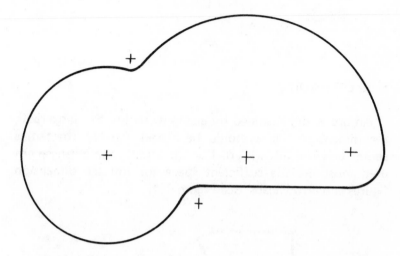

PROBLEM 3–1. Size of radii. Remove and tape down the duplicate worksheet at the back of the book. Scale the drawing to determine sizes and dimension the sizes of the radii according to the examples shown in **Figure 3–1.**

Repeated Radii. Where a part has a number of radii of the same size, a note may be used instead of dimensioning each radius separately. Elaborate leader systems are undesirable. Making the drawing user assume any of the dimensions can cause confusion (**Figure 3-2**).

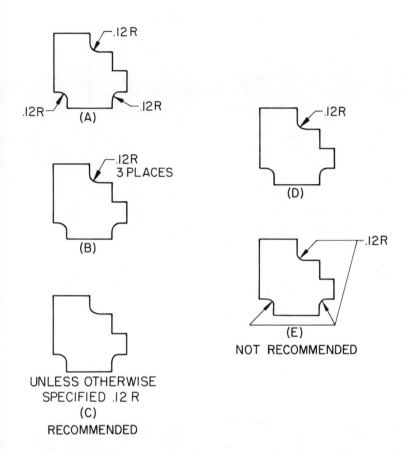

Figure 3-2. Repeated radii. **A**, **B** and **C** show three acceptable ways of dimensioning repeated radii. **D** should be avoided on complicated drawings because forcing the user of the drawing to make assumptions about undimensioned radii could cause errors. **E** should be avoided because elaborate leader line systems can be confusing.

True Radii. Where the radius is dimensioned in a view that does not show the true shape, **TRUE R** is added after the radius dimension.

Shortened Radii. Where the center of a radius is outside the drawing or interferes with another view, the radius dimension line can be shortened. See **Figure 3-3.**

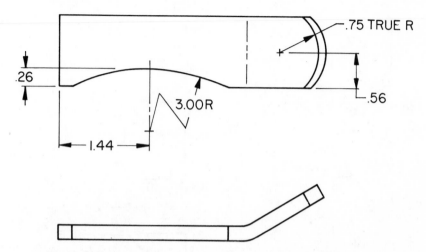

Figure 3-3. True radius and shortened radius dimensioning. Since the end of the strap is bent up, the true shape of the .75 R would only be seen in an auxiliary view and the word "TRUE" must be added if the arc is to be dimensioned in a principal view. The shortened radius approach is used on the **3.00 R** arc because the true center interferes with the bottom view.

Spherical Radii. Where spherical surfaces are dimensioned, **SPHER R** is added after the radius size. See **Figure 3-4.**

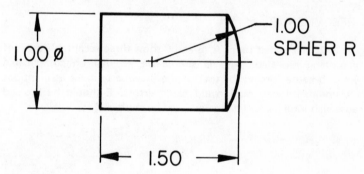

Figure 3-4. Spherical radius. The shaft has a ball shape on the end.

LOCATION OF ARCS

An arc may be located by tangent lines or centerlines. Where the location of the center is unimportant, tangent lines are used to locate the arc and the center is not shown. Where the location of the center is important (as in cases where two or more arcs have the same center) the center is located. See **Figure 3-5.**

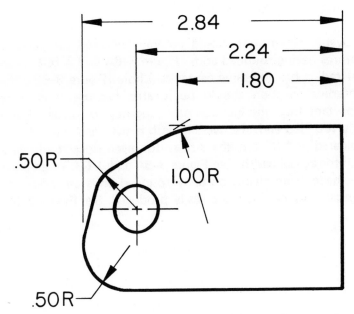

Figure 3-5. Location of radii. While two of the arcs shown are located by tangent lines, the third is located by centerlines because the center of the radius is also the center of the hole.

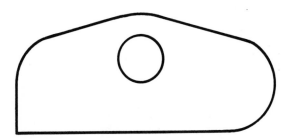

PROBLEM 3-2. Location of radii. Remove and tape down the duplicate worksheet at the back of the book. Dimension the location of the radii according to the examples shown in **Figure 3-5** and the size of the radii according to previous examples. To determine the sizes, use your circle template. (Remember that the sizes on your circle template are diameters and the dimensions you are trying to find are radii).

ROUNDED—END PARTS

Overall dimensions should generally be used for parts or features having rounded ends (**Figure 3-6A** and **3-6C**). If an end radius has the same center as a hole (**Figure 3-6B**), the centerline location should be located because it is more important than the location of a tangency to the arc. For a fully rounded end, the overall width is given and the radius is indicated as "R" but the size is not given since it will vary with the actual width. See **Figure 3-6A** and **3-6B**. When parts are made from strips, partially rounded ends are preferred because they can be more easily produced. See **Figure 3-6C**.

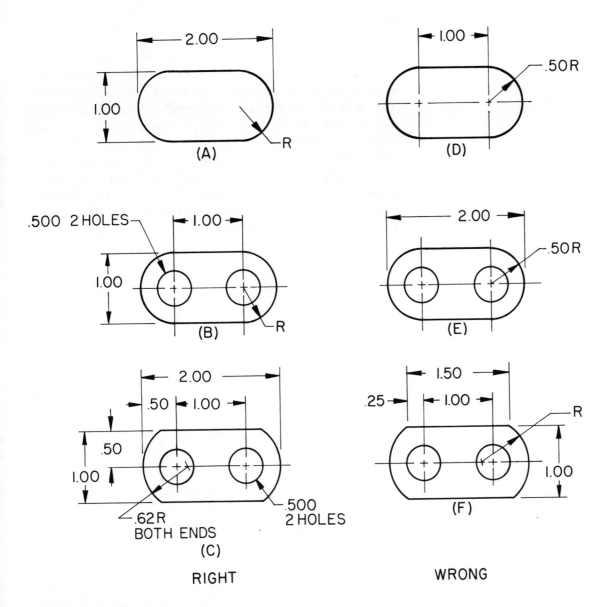

RIGHT   WRONG

Figure 3-6. Rounded-end parts. **A** is preferable to **D** because the overall length and width in **A** are more important for the manufacture and the proper functioning of the part than the distance between centers in **D**. **B** is preferable to **E** because the center distance between the holes in **B** is more important for the proper functioning of the part than the overall length in **E**. **C** is preferable to **F** because the dimensions in **C** contain all the information necessary to produce the part while **F** contains some irrelevant information and is lacking some necessary information.

SLOTTED HOLES

Most slots are elongated holes which are punched in sheet metal and used instead of conventional holes to compensate for manufacturing inaccuracies or to provide for adjustments (**Figure 3-7**). The size may be specified with a note which gives the overall length and width, or it may be specified with length and width dimensions. If the center point of one or both ends of the slot is also the center of a hole or another arc, the length of the slot may be dimensioned to the center rather than to the end as shown in **Figure 3-6B**. Slots are located by dimensions to the centerlines or to the longitudinal centerline and one end.

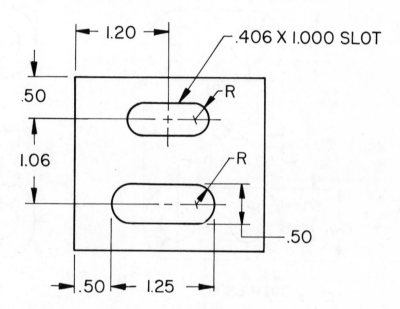

Figure 3-7. Slotted holes. The overall length and width of the upper slot is dimensioned by note and the lower slot by dimensions. The upper slot is located by dimensions to the centerlines and the lower slot by dimensions to the longitudinal centerline and one end.

PROBLEM 3-3. Rounded-end parts. Remove and tape down the duplicate worksheets at the back of the book. Dimension each object completely according to the scaled sizes.

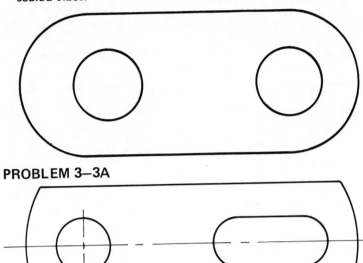

PROBLEM 3—3A

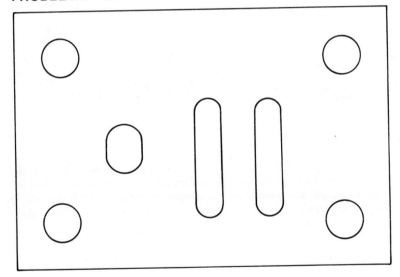

PROBLEM 3—3B

PROBLEM 3—3C

ROUNDED CORNERS

Where corners are rounded, dimensions should locate the horizontal and vertical edges and the arcs will be located by being tangent to the located edges. See **Figure 3-8.**

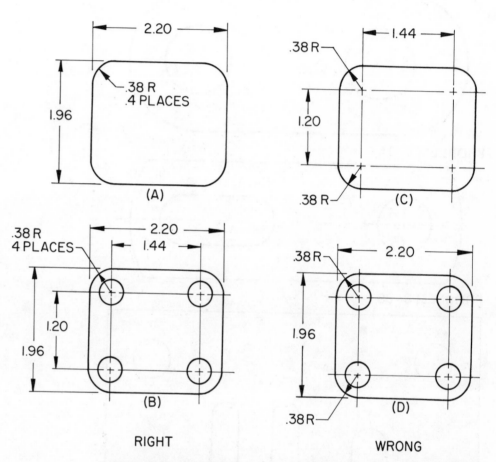

RIGHT   WRONG

Figure 3-8. Rounded-corner parts. **A** is preferable to **C** because the vertical and horizontal sides of the object should be located in preference to the unimportant radius centers. **B** is preferable to **D** because the relationship between the holes is just as important as the overall length and width of the part.

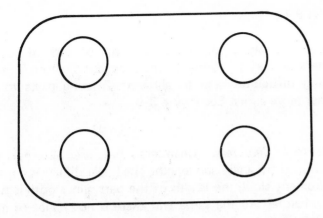

PROBLEM 3-4. Rounded–corner parts. Remove and tape down the duplicate worksheet at the back of the book. Dimension the object completely.

3-2 FEATURES ON CYLINDERS

Machine parts are often made from flat or round bar stock which is produced by passing white-hot slabs of steel between rollers. This hot rolling is followed by cold working or cold finishing when a better surface finish, greater dimensional accuracy, or improved mechanical properties are required.

Steel shafts used on machine parts are either hot-rolled or cold-finished round bars which are cut off and made into a usable shape by a lathe and other machine tools. The commonly specified features on shafts are chamfers, grooves and threads.

CHAMFERS

These are bevels or angles cut on corners of round cylinders or bars to eliminate sharp edges which may cause handling difficulties and to allow the mating parts to fit together more easily. See **Figure 3-9**.

External Chamfers. Chamfers that are external are dimensioned by angle and length. The linear dimension is the measurement along the length of the part. Since both legs of a 45° chamfer are the same and there is no chance of misinterpretation, chamfers of that size may be specified by note. When the chamfer joins two portions of shaft, either the angle or the length dimension must be omitted to avoid over-dimensioning.

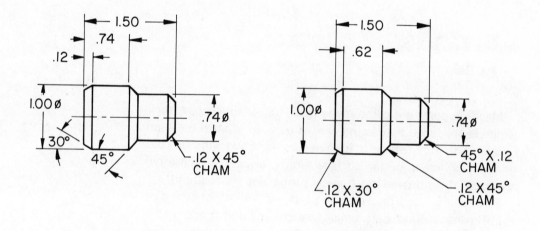

RIGHT   WRONG

Figure 3-9. External chamfers. The dimensioning of the left end chamfer of the "right" example is preferable to the dimensioning of the "wrong" example because only 45° chamfers may be dimensioned by note. The dimensioning of the center chamfer or taper of the "right" example is preferable to the "wrong" example because the .12 leg size in the latter is not needed. The dimensioning of the right end chamfer of the "right" example is preferable to the "wrong" example because the chamfer leg must be listed before the chamfer angle.

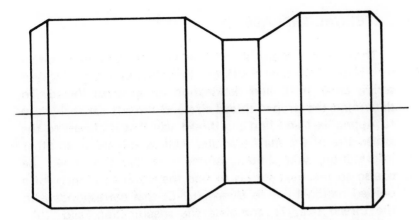

PROBLEM 3-5. External chamfers. Remove and tape down the duplicate worksheet at the back of the book. Dimension the chamfers according to the examples shown in **Figure 3-9.**

Internal Chamfers. When a chamfered hole is over one inch in diameter, a general-purpose tool is used to produce the chamfer and the approach to dimensioning is the same as with an external chamfer. If the chamfered hole is under one inch or in any other case where the chamfer diameter and included angle are most important, the countersink approach to dimensioning is preferable. See **Figure 3-10.**

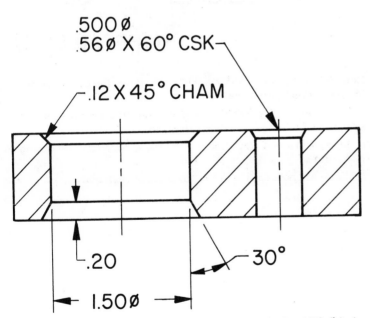

Figure 3-10. Internal chamfers. The chamfers on the 1.50 Ø hole are dimensioned like external chamfers and the chamfer on the .500 Ø hole is dimensioned as a countersink.

EXTERNAL THREADS

Threads on the outside of a cylinder or shaft are specified by a note which gives the thread size, threads per inch, thread series, class of fit and designation for external thread. To determine the proper thread size and threads per inch, turn to Appendix I and find the thread size that is closest to the scaled size of the shaft diameter. Unless you are given other information, you should assume that the thread will be coarse so the next step is to find the threads per inch for a unified national coarse thread (UNC) that corresponds with the thread size. To complete the specification, add "2A" which means that it is a general purpose, external thread. See **Figure 3–11.** The leader from the note extends to the rectangular view of the threaded shaft.

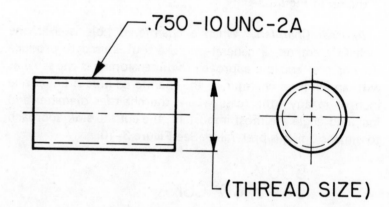

Figure 3–11. External threads. Here are three things to remember when specifying external threads:

(1) The leader for the thread note is directed to the rectangular rather than the circular view of the external thread.

(2) No tap drill size is given since no drilling is involved in producing an external thread.

(3) The thread note ends with "A" to designate external.

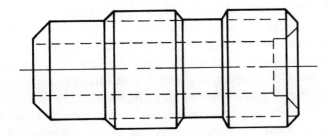

PROBLEM 3-6. External thread specifications. Remove and tape down the duplicate worksheet at the back of the book. This part is a hydraulic fitting with threads on portions of the outside and a hole going through the center. Dimension the part completely. Give the external thread specifications according to the examples shown in **Figure 3-11**. Determine the thread size by scaling the major diameter and find the thread specification data in Appendix I at the back of the book.

GROOVES

Grooves are often machined at shoulders on shafts or bores for ease of machining and assembly of parts (**Figure 3-12**). Although most companies have their own variation, the two basic types are round- and flat-bottomed grooves.

Round-Bottomed Grooves. These grooves are often used to allow a mating part to slide up to a shoulder on a shaft. A sharp corner would allow assembly but would cause a stress concentration, making the part prone to crack at the corner. A radius spreads out the stress and is less apt to create a breaking point. The reliefs used for this purpose are usually dimensioned by diameter of the groove and radius. Some versions also include an angle as an aid to machining.

Flat-Bottomed Grooves. These grooves are often used as thread reliefs to compensate for the fact that it is not possible to make a complete thread up to a shoulder. Without a relief of the thread at the shoulder, assembly problems result since this condition prevents a nut from being threaded up against the shoulder. Flat-bottomed reliefs are dimensioned by the diameter, width and inside corner radii. The diameter must be smaller than the minor diameter of the thread as found in *Machinery's Handbook* or some other reference providing a comprehensive listing of thread information. The width of the groove must be equal to at least three threads to eliminate incomplete threads. The inside corner radii are commonly .03 R, just big enough to prevent stress concentrations.

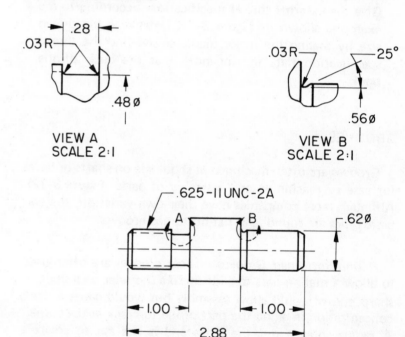

Figure 3–12. Grooves. The use of blown-up views clarifies small details. Be sure to remember that dimensions must match the actual part and not the enlarged scale picture. View **A** shows a flat-bottomed groove used for a relief for the **.625–11UNC** thread. View **B** shows a round-bottomed groove used as an aid for machining and assembly of parts.

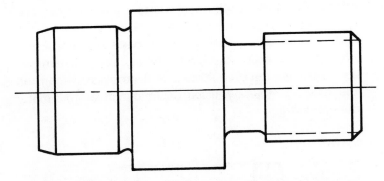

PROBLEM 3-7. Grooves. Remove and tape down the duplicate worksheet at the back of the book. Dimension the part completely using blown-up views to show the grooves. Follow the example shown in **Figure 3-12.**

Knurls. Knurls represent a texture or design which is used to roughen a surface for better grip, for decoration or for press fits between parts. Specifications always include pitch, type of knurl, diameter before knurling and the length of the area to be knurled. The standard types of knurls are straight, diagonal and diamond. For additional information on standard pitches and tolerances in connection with knurling, see the *ANSI B94.6 Standard* or the *Machinery's Handbook.*

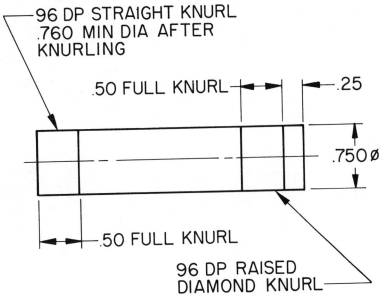

Figure 3-13. Knurls. **96 DP** (diametrical pitch) is the pitch most commonly used in knurls. A 96 DP knurl contains 96 serrations, or grooves, on the circumference for each inch of shaft diameter. Since the shaft is .750 Ø, each knurl in the example contains 72 grooves (96 X .75).

PROBLEM 3-8. Cylindrical parts dimensioning. Remove and tape down the duplicate worksheets at the back of the book. Each of these parts is a shaft into which have been machined holes, radii, chamfers, countersinks, internal threads and grooves. Dimension the parts completely using decimal-inch or millimeter dimensioning.

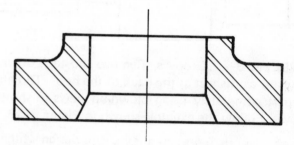

PROBLEM 3—8A

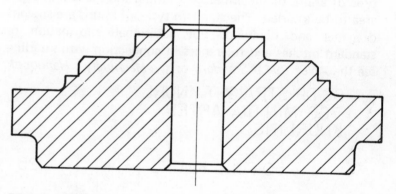

PROBLEM 3—8B

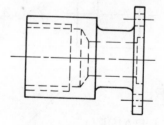

PROBLEM 3—8C

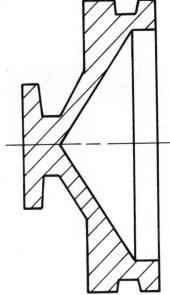

PROBLEM 3–8D

3-3 LEARNING APPLICATION

The drawings required for Problem 3-9 will show how well you have met the module objectives.

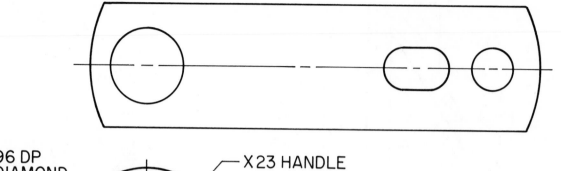

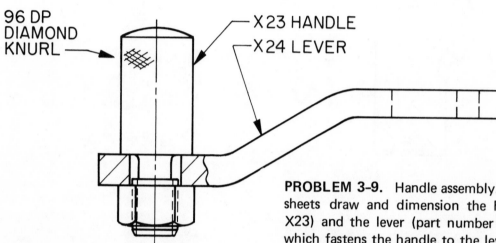

96 DP
DIAMOND
KNURL

X23 HANDLE
X24 LEVER

PROBLEM 3-9. Handle assembly. On separate drawing sheets draw and dimension the handle (part number X23) and the lever (part number X24). Since the nut which fastens the handle to the lever is a standard part, it is not detailed.

DRAWING CHECKLIST

1. Are the radius dimensions arranged so that arrowheads and dimension figures are full-sized without being cramped?
2. Do the radius dimension lines extend from the radius centers at an angle?
3. Are overall dimensions used on rounded-end parts?
4. Are the radii located by tangencies when the centers are unimportant or when the edges are vertical or horizontal lines?
5. Are the 45° chamfers dimensioned by note and others by length and angle?
6. Are external threads dimensioned in the rectangular view and given the external thread designation?
7. Are the round-bottomed grooves dimensioned by diameter for the depth and by radius for the width?
8. Are the flat-bottomed grooves dimensioned with a groove diameter that is smaller than the root diameter of the thread, a width that is greater than three threads, and with small corner radii?
9. Are the knurls dimensioned by pitch, type, diameter, length and location?

ANSWERS TO PROBLEMS

These partial solutions contain some of the dimensions and a general approach which may be used. Other approaches may also satisfy the ANSI dimensioning requirements.

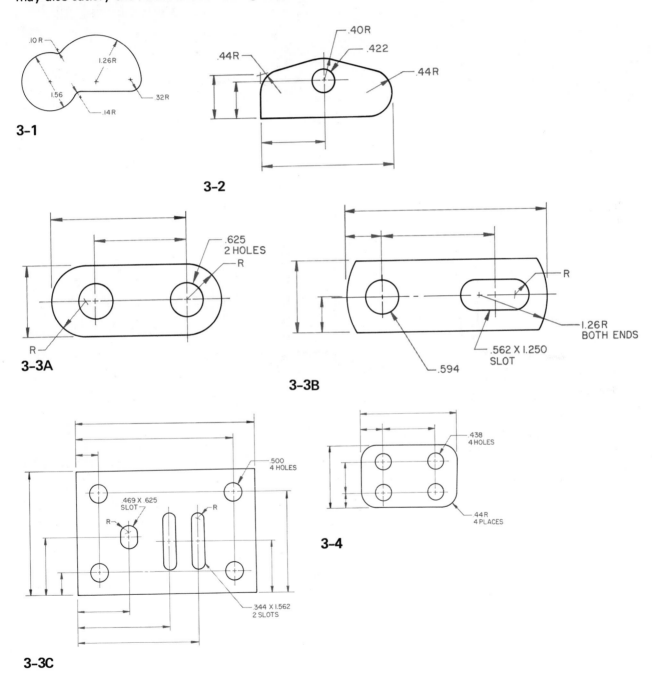

3-1

3-2

3-3A

3-3B

3-3C

3-4

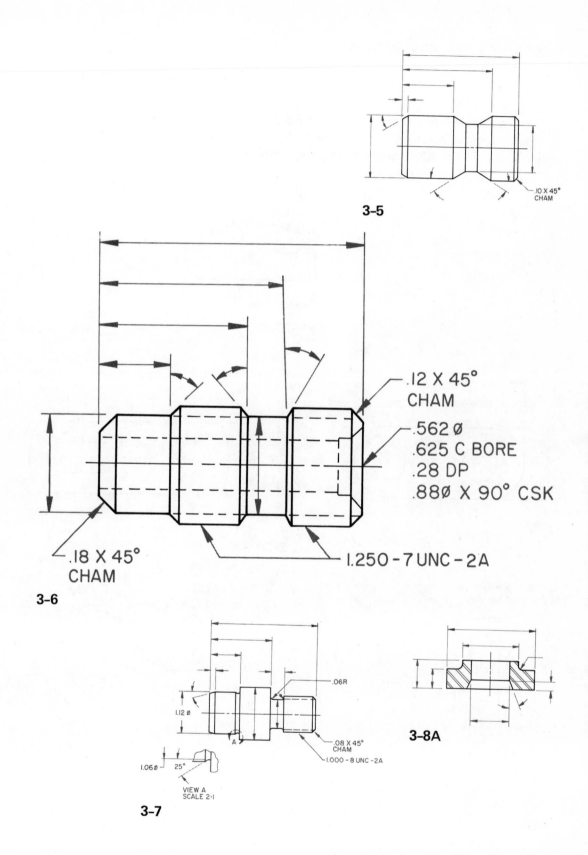

3-5

.10 X 45°
CHAM

.12 X 45°
CHAM

.562 Ø
.625 C BORE
.28 DP
.880 X 90° CSK

.18 X 45°
CHAM

1.250 - 7 UNC - 2A

3-6

.06R

1.12 Ø

A

.08 X 45°
CHAM

1.000 - 8 UNC - 2A

1.06 Ø   25°

VIEW A
SCALE 2:1

3-7

3-8A

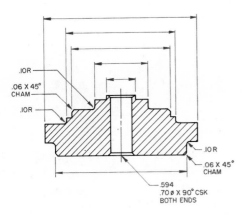

.IO R
.06 X 45°
CHAM
.IO R

.IO R
.06 X 45°
CHAM

.594
.70 ⌀ X 90° CSK
BOTH ENDS

3–8B

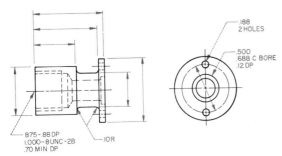

.188
2 HOLES

.500
.688 C BORE
.12 DP

.875 - .88 DP
1.000 - 8 UNC - 2B
.70 MIN DP

.IO R

3–8C

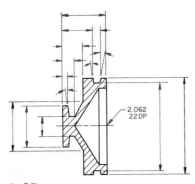

2.062
22 DP

3–8D

4

COMPLETENESS

It is essential that a drawing have just the right amount of dimensioning. If some dimensions are missing, and are not directly calculable, the users of the drawing cannot be sure what sizes were intended. If there are too many dimensions, the users of the drawing cannot be sure which dimensions they are to work from in order that the mating parts and tolerances be met.

PURPOSE

The purpose of this module is to show you how to determine the quantity of dimensions to use in order to show clearly the intent of the drawing.

PREREQUISITES

Before you begin this module, you should be able to:

1. Apply basic drafting skills including construction of tangent radii (Worktext 1).
2. Apply ANSI standards for dimensioning common features (Modules 1 through 3).

OBJECTIVES

Upon completion of this module, you will be able to:

1. Dimension parts completely enough so that the intent is clear.
2. Dimension symmetrical features, including cases in which overall widths are more critical than the dimensions to the axis of symmetry.
3. Dimension lengths and angles without double dimensioning.
4. Dimension tangent radii completely enough for construction.

EQUIPMENT AND REFERENCES

Before you begin this module, you should have:

1. All basic drafting equipment including compass.
2. ANSI Y14.5, *Dimensioning and Tolerancing.*

4-1 ASSUMING DIMENSIONS

"Dimensions for size, form and location of features must be complete to the extent there is full understanding of the characteristics of each feature. Neither scaling nor assumption of a distance or size is permitted." (ANSI Y14.5, Paragraph 1.7b).

The purpose of this rule is to prohibit under-dimensioning so that the mistakes and confusion that result from scaling a drawing or making assumptions to determine dimensions can be avoided. Full understanding of the intent of a drawing requires completeness in dimensioning. See **Figure 4-1.**

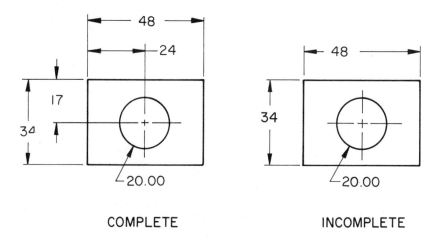

COMPLETE   INCOMPLETE

Figure 4-1. Complete and incomplete dimensioning. In the "incomplete" example, the user of the drawing is forced to make assumptions about the location dimensions and tolerances.

TYPICAL DIMENSIONS

In the case of repeated features on simple parts, full understanding is often possible if one of a number of obviously identical features is dimensioned completely and the same dimensions are presumed to apply to the undimensioned features as well. The resulting reduction in the total number of dimensions on the drawing makes a drawing that is actually more easily understood than one in which every feature is dimensioned (**Figure 4-2**).

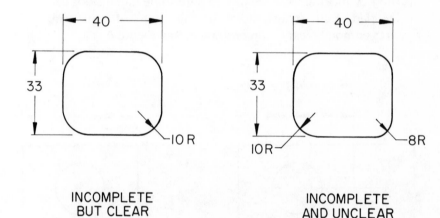

INCOMPLETE
BUT CLEAR

INCOMPLETE
AND UNCLEAR

Figure 4-2. Typical dimensions. Since all the radii are the same in the figure on the left, one dimension is sufficient. Since there are two sizes in the figure on the right, the sizes of the upper arcs are not clear and they need to be dimensioned as well.

DIMENSIONING SYMMETRICAL FEATURES

Dimensioning symmetrical features may be simplified by identifying the axis of symmetry and dimensioning from the axis to features on one side only (**Figure 4-3**). The dimensions would then apply to both sides of the axis. However, for symmetrical features like mating tongue and grooves and for flat stock with holes in the center, the overall widths are more important than the distance from the side to the axis of symmetry. If the overall distance alone was given, the distance from the side to the center could be assumed to be half of the width, but the inspector for the parts would have no way of knowing the tolerance for the center locations. Such a tolerance could be the same as the standard machining tolerance, the same as the tolerance for the width, or an amount half of the tolerance for the width. To avoid misunderstanding, both the overall width and the dimension from the side should be given. The required tolerances, or a note, should be given to tell how far off center the axis of symmetry could be. (An exception to this would be cylindrical objects since the dimension from the center to the side should never be given for a full diameter.)

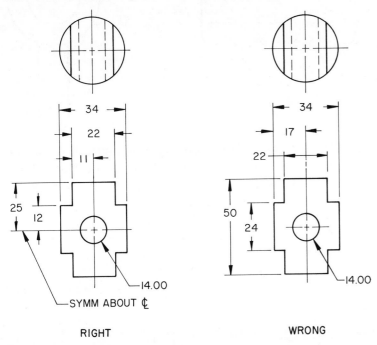

RIGHT   WRONG

Figure 4-3. Dimensioning symmetrical features. Since 34 is the diameter of a cylinder, the 17 dimension from the side of the cylinder to the cylinder centerline in the "wrong" example is incorrect. However, 22 is the width of a non-cylindrical feature, so the 11 dimension from its side to the cylinder centerline is necessary. Since the 50 dimension is the size of a non-cylindrical feature, both the 14.00 diameter hole and the 24 wide feature need to be located in some way rather than be assumed as centered as shown in the "wrong" example.

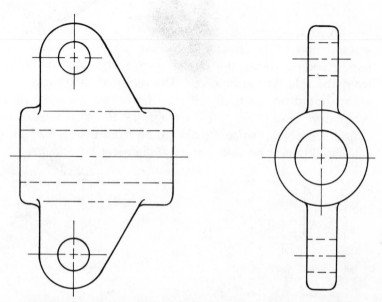

PROBLEM 4-1. Dimensioning symmetrical features. Remove and tape down the duplicate worksheet at the back of the book. Dimension completely in millimeters. Bear in mind that the distance from a feature to the centerline must be exactly half of the desired distance to the same feature on the other side of the axis of symmetry.

4-2 DOUBLE DIMENSIONING

REPEATED DIMENSIONS

A given dimension should be shown only once on the drawing (**Figure 4-4**). Although the dimension may be easier to find if it is shown more than once, the drawing could become less clear because of the greater quantity of dimensions, and there could be more chance of errors and conflicting dimensions. If a repeated dimension will add greatly to the clarity of a drawing, it can be shown in parentheses as a *reference dimension*, a non-toleranced dimension used for clarification but not for performance of work or inspection.

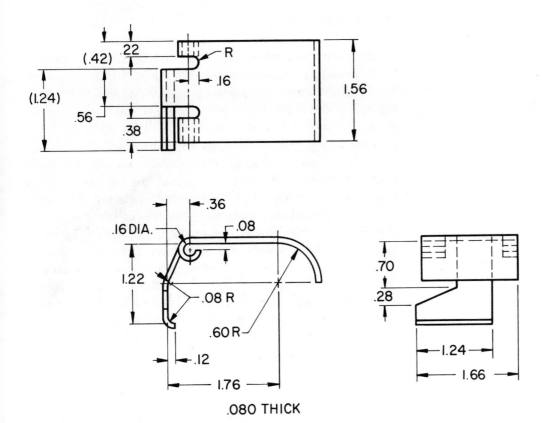

.080 THICK

Figure 4-4. Reference dimensions. The 1.24 dimension in the top view is shown in parentheses to indicate that it is a reference dimension because it is also shown in the side view. The .42 reference dimension in the top view is not shown in another view, but it could only be shown as a reference dimension because it may be found by finding the difference between the 1.66 and 1.24 dimensions in the side view. In both cases, the reference dimensions make the top view easier to understand.

OVER-DIMENSIONING

While the "Completeness Rule" (ANSI Y14.5, Paragraph 1.7b) prohibits under-dimensioning, ANSI Y14.5, Paragraph 1.7d prohibits over-dimensioning or any practice that would result in more than one interpretation. If a size or location is given in more than one way, there will be more than one possible interpretation. If parts could be made to the exact size, this type of double dimensioning would not be a problem because the machinist would get the same results regardless of which of the dimensions he used. However, since the dimensions cannot be held with complete accuracy, and since the inaccuracies accumulate, a part must be made to one set of dimensions and tolerances. Otherwise, the part could meet the first set of dimensions and tolerances, only to be scrapped because of failure to meet another set of dimensions and tolerances. To arrive at a single interpretation set only, a part must have a minimum number of dimensions that will allow the part to be produced without confusion, conflicts or assumptions.

LENGTHS

If a part has a complete chain of dimensions and an overall length, the least important dimension, usually one of the intermediate lengths, should be omitted to avoid double-dimensioning (**Figure 4-5**). This approach will provide enough information for the part to be produced without making any assumptions. If necessary for clarity, the extra dimension may be given as a reference dimension.

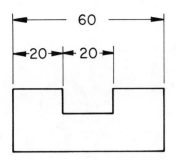
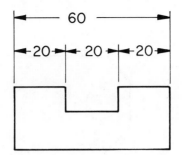

RIGHT   WRONG

Figure 4-5. Lengths. In the "right" example, each vertical surface is located by one and only one dimension and will be affected by one and only one tolerance for one interpretation. In the "wrong" example, there is both a complete chain of dimensions and an overall length which means that one of the surfaces would be affected by conflicting tolerances for two possible interpretations.

ANGLES

An angle or taper may be dimensioned by two linear dimensions, or one linear and one angular dimension (**Figure 4-6**). The choice should depend on the function of the feature and its relationship to the object and mating objects.

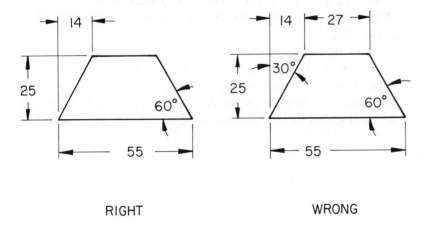

RIGHT   WRONG

Figure 4-6. Angles. In the "right" example, every dimension is essential, while in the "wrong" example, the part could be constructed and produced even if the 14 and 27 dimensions were removed.

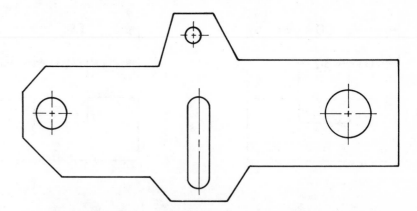

PROBLEM 4-2. Lengths and angles. Remove and tape down the duplicate worksheet at the back of the book. Dimension completely in millimeters and degrees. On any drawing such as this it is recommended that you tie all the holes together and give as many overall sizes as possible before dimensioning the angled features.

RADII

Radii are also dimensioned on the basis of the minimum amount of information needed to construct them. Remember that when the shop worker lays out the radii, he applies the same geometric construction principles that the draftsperson must apply. See **Figure 4-7** and **Figure 4-8**. (Refer to Worktext 1 for a review of geometric construction principles).

If an arc is to be tangent to two located features and one centerline of the arc is located, the arc may be constructed even if its size is not known. In this case the size of the arc is given as a reference dimension or the radius is indicated without giving its size. See **Figure 4-8.**

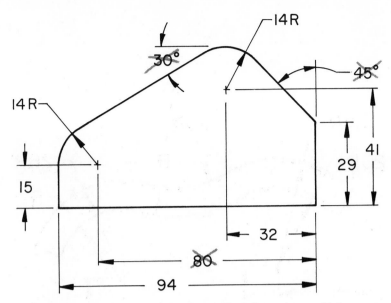

Figure 4-7. Arcs tangent to lines. The dimensions which are not needed to construct the object are unnecessary and have been crossed out.

PROBLEM 4-3. Arcs tangent to lines.

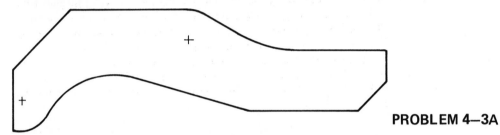

PROBLEM 4—3A

(A) Remove and tape down the duplicate worksheets at the back of the book. Dimension completely.

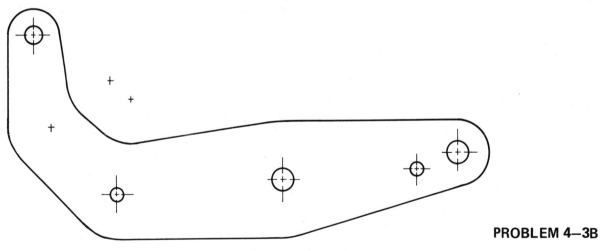

PROBLEM 4—3B

(B) Check your dimensioning completeness by using your dimensions to lay out the objects on another drawing sheet. Change your dimensions if necessary.

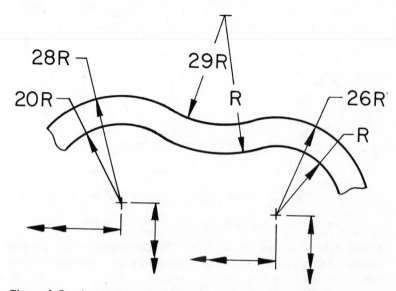

Figure 4–8. Arcs tangent to arcs. The double arrows on the dimension lines are a way of indicating that the features are being located from unseen reference lines. The centers of the 28 and 26 radii are located by dimensions to the vertical and horizontal centerlines and the center of the 29 R can be located by geometric construction so it must not be located with dimensions. Once the 20 R arc is swung from the center of the 28 R arc, the undimensioned "R" arcs may be swung from the other two centers to complete the construction even though the arc sizes are unknown.

PROBLEM 4-4. Arcs tangent to arcs.

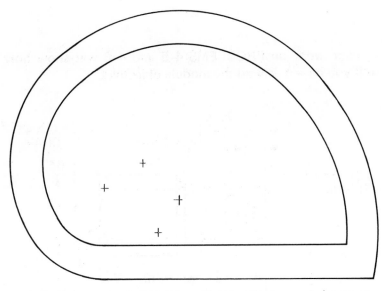

PROBLEM 4–4A

(A) Remove and tape down the duplicate worksheets at the back of the book. Dimension the gaskets completely.

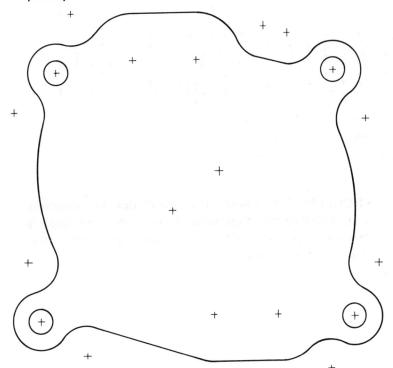

PROBLEM 4–4B

(B) Check your dimensioning completeness as in Problem 4-3.

Your work on PROBLEMS 4-5 and 4-6 will show how well you have mastered the module objectives.

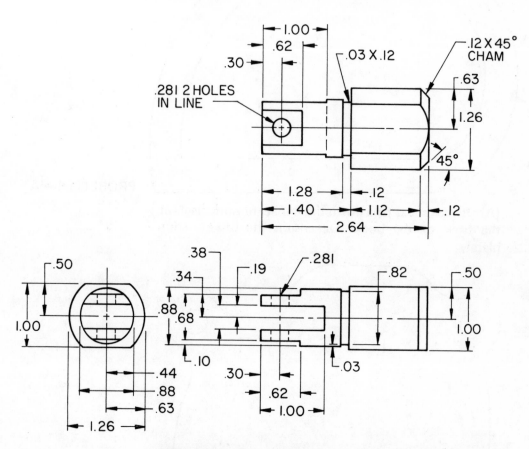

PROBLEM 4-5. Lever. The given drawing contains both desirable and undesirable dimensions. Tape down a drawing sheet and draw the lever, using only the desirable dimensions.

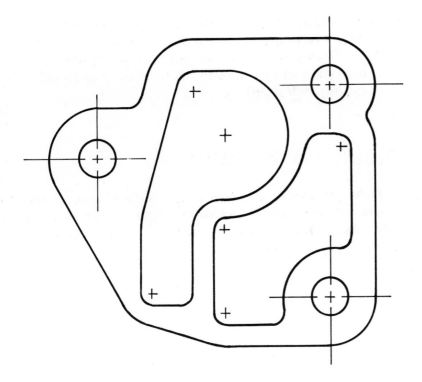

PROBLEM 4-6. Gasket. Remove and tape down the duplicate worksheet at the back of the book. Dimension the gasket in millimeters. Then, tape down a drawing sheet and redraw it using your dimensions.

DRAWING CHECKLIST

1. Is dimensioning complete enough for the shop to make the part just as it was intended?
2. Have typical dimensions been used only in cases where there could be no misinterpretation?
3. Have full diameters been dimensioned as diameters rather than radii?
4. Have symmetrical features been dimensioned to a labeled axis of symmetry on one side only in order to simplify dimensioning of parts in which the overall dimensions were not critical?
5. Where overall widths of symmetrical features were important, was the width and the distance from the side to the axis of symmetry given?
6. In a chain of dimensions, has either the overall length or one of the intermediate lengths been omitted?

7. In dimensioning an angle, has either the angular dimension or one of the legs been omitted?
8. In dimensioning radii, has just enough information been given so that the size and location of each radii can be constructed?

ANSWERS TO PROBLEMS

These are partial solutions which contain some of the dimensions and a general approach which may be used. In some cases other approaches may also satisfy the ANSI dimensioning requirements.

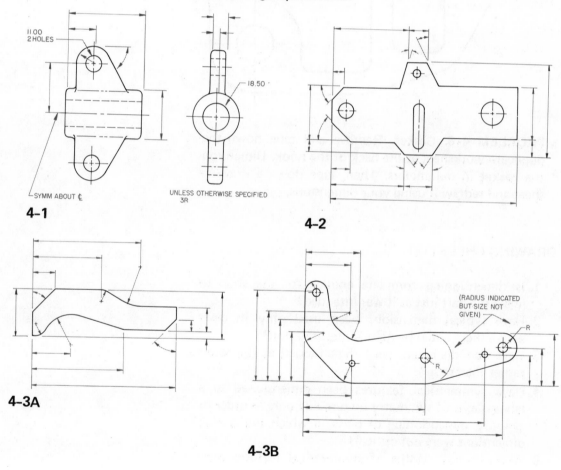

4-1

4-2

4-3A

4-3B

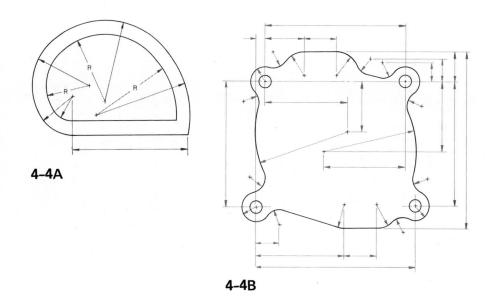

4–4A

4–4B

5

CASTINGS

While the ANSI dimensioning rules do not permit dimensioning to dictate a particular manufacturing process, it is necessary for the draftsperson to be familiar with the processes so that the dimensioning system used does not make the part unnecessarily difficult to make when the most economical manufacturing method is used.

PURPOSE

The purpose of this module is to show you how to dimension a machined or non-machined casting in a way that takes into consideration both the ANSI dimensioning rules and the casting process.

PREREQUISITES

Before you begin this module, you should be able to:

1. Apply basic drafting skills and multiview drawing principles (Worktexts 1 and 2).
2. Apply the ANSI Y14.5 standards in dimensioning common shapes and features (Modules 1 through 4).

OBJECTIVES

Upon completion of this module, you will be able to:

1. Dimension draft, fillets and rounds on a casting drawing.
2. Identify machined surfaces on a machined casting drawing with the use of finish marks.
3. Dimension a machined casting drawing with a minimum of finish-to-rough dimensions.
4. Dimension cast and machined surfaces on the same drawing, specifying casting and machining toleranced dimensions.

EQUIPMENT AND REFERENCES

Before you begin this module, you should have:

1. Typical drafting tools, including lead holder, drafting machine, decimal-inch and metric scales and circle template.
2. ANSI Y14.5, *Dimensioning and Tolerancing.*

5-1 CASTING DRAWINGS

DRAFT ANGLES

Draft angles are built into the sides of a pattern so that the pattern may be removed from the sand in the mold without disturbing any detail in the mold. The draft angle starts at the parting line, that is, the line that separates the portion of the pattern that is in the *cope* (the top half) from the portion in the *drag* (the bottom half).

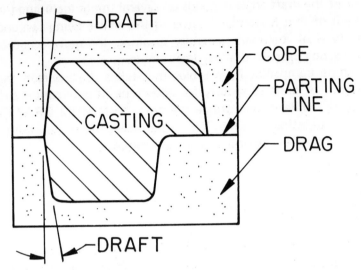

Figure 5-1. Draft angles and parting line. This shows the mold after the casting has been poured. The draft angles permit the cope and drag halves to be separated and the pattern removed without disturbing the sand in the mold.

In many cases, the draft angle is not shown on a casting drawing because it is too small to make any difference in the dimensioning of the part. When it is necessary to show the draft because of the shape or function of the part, the dimensions should be given to either the top or bottom of the draft, whichever is more important to the function and production of the part.

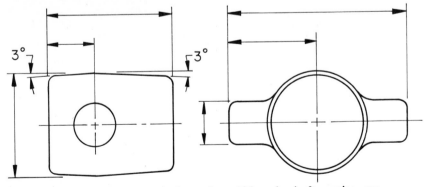

Figure 5-2. Dimensioning draft angles. Although draft angles are usually specified in a general note, in this case the drawing is clearer with the angles specified by dimensions.

Specification of Draft. While it is sometimes necessary to dimension draft angles individually, it is usually more practical to specify them in a general note. Although 2° draft is specified on most examples in this module, this is not a typical draft angle that can be specified for all castings. The size of the draft angle depends on several things including the length of that particular feature of the article being cast and the type of material. Generally speaking, the more accurate the casting process, the less draft required.

Some companies have tables that tell a draftsperson the amount of draft to specify for each type of application and process, while others utilize the note "MINIMUM DRAFT" on each casting drawing. This note places the responsibility for draft angle selection on the patternmaker.

FILLETS AND ROUNDS

One of the most noticeable characteristics of castings is that most of the internal and external corners are rounded. The inside radii are *fillets* and the outside radii are *rounds.* They aid in the casting process and reduce stress concentrations in the castings.

Specification of Fillets and Rounds. The specification of these radii can be done by local notes on the drawing, but when there are several fillets or rounds of the same size, the use of a general note is the simplest and clearest approach.

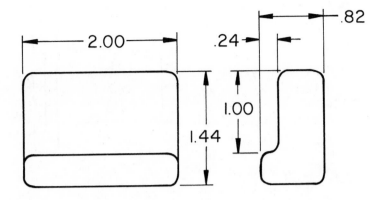

1. UNLESS OTHERWISE SPECIFIED
    2° DRAFT
    .12 R
    ±.04 TOLERANCES
2. MATERIAL ASTM A4864
        GRAY IRON

PART NO. X238

Figure 5-3. Casting drawing. Casting dimensions require a larger tolerance than is normally applied to machining drawings. This drawing would be used by the pattern shop and foundry. General notes include draft, radii and material.

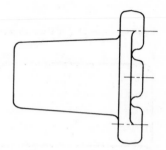

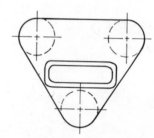

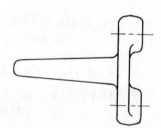

PROBLEM 5-1. Casting drawing. Remove and tape down the duplicate worksheet at the back of the book. Dimension the casting completely, using general notes to specify draft (3°) fillets and rounds.

5-2 MACHINED CASTING DRAWINGS

Since casting is a relatively inaccurate process, most castings require some machining to be usable. A casting to be machined may be drawn as two separate casting and machining drawings, or together as a single casting/machining drawing, depending on company practice. **Figure 5-3** is an example of a casting drawing and **Figure 5-4** is a machining drawing of the same part. A casting/machining drawing of the part is shown in **Figure 5-5**.

MACHINING DRAWINGS

When there are two separate drawings for casting and machining, the part number for the casting drawing is listed as the material for the machining drawing, and the casting dimensions are not repeated. The pattern shop and the foundry perform their work from the casting drawing, while the machine shop works from the machining drawing.

Locating Machined Surfaces. While the machinist always needs a few finish-to-rough dimensions to provide a starting place for machining, such dimensions should be kept to a minimum. Since the most important relationships are usually between the machined surfaces, and since accurate measurements cannot be made from the rough cast surfaces, finish-to-finish dimensions are preferred over finish-to-rough dimensions.

Finish Marks. These marks are symbols that are used on machining drawings to tell the machinist which surfaces are to be machined. They are drawn on the "air" side of the edge views of machined surfaces, that is, the side where the rough metal is to be removed.

There are several types of finish marks, and some companies show them on all machined edges while others show them only in views in which a dimension goes to the edge. Since design layouts do not show finish marks, the draftsperson must determine where they go by looking at the inside and outside corners of the casting. Rounded corners are usually cast and sharp corners usually represent cases in which one or both of the intersecting surfaces are machined. The only sharp cast corners are those that take place in some cored holes or at the parting line as shown in **Figure 5–1**. A *cored hole* is the term used for a hole that is cast into the part, since a core is the device used to form the hole. Most holes over one inch are cored.

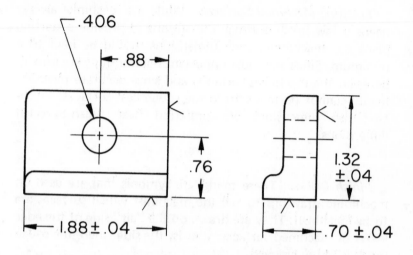

1. MATERIAL: PART NO. X238
2. UNLESS OTHERWISE SPECIFIED
   ±.02 TOLERANCES

PART NO. X239

Figure 5-4. Machining drawing. This is the drawing the machinists would refer to when performing the machining operations on the casting (Part no. X238) shown in **Figure 5-3**. Notice that none of the casting dimensions or notes shown in **Figure 5-3** are repeated on this drawing. Finish-to-rough dimensions require larger tolerances than finish-to-finish dimensions. The check marks shown on the machined surfaces are finish marks. While part number X238 is specified as the material, ASTM A4864 could be included as a reference in the material specification.

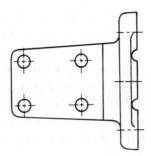

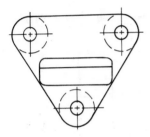

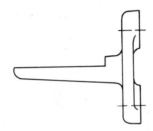

PROBLEM 5-2. Machining drawing. This is the machined version of the part you dimensioned in Problem 5-1. Remove and tape down the duplicate worksheet at the back of the book. Dimension the machined surfaces completely, using a minimum of finish-to-rough dimensions to relate the machined surfaces. Do not repeat any of the casting dimensions used in Problem 5-1. Apply tolerances according to the example in **Figure 5-4.**

CASTING/MACHINING DRAWINGS

These are multiple-purpose drawings used by the pattern shop, foundry and machine shop to make a machined casting from start to finish. Once again, a few finish-to-rough dimensions are necessary, but they should be kept to a minimum. Rough-to-rough dimensions are good because they can be used directly by the pattern shop and foundry. Finish-to-finish dimensions are good because they provide the most accurate relationship between the most important features.

Finish. While a casting/machining drawing needs to have specifications for material, draft and radii just like a casting drawing, one additional item is required. A general note reading ".12 FINISH" tells the patternmaker that an extra .12 of material needs to be added to all surfaces that are indicated by finish marks to be machined surfaces. While the amount of finish may vary somewhat with different applications and processes, .12 finish is very typical for gray iron castings.

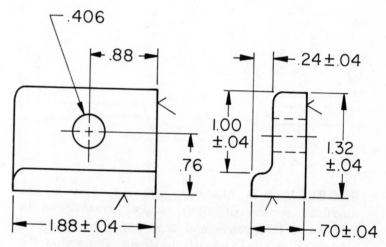

1. UNLESS OTHERWISE SPECIFIED
   .12 FINISH
   2° DRAFT
   .12 R

2. MATERIAL: ASTM A4864
   GRAY IRON

Figure 5-5. Casting/machining drawing. This version of the part shown in **Figure 5-3** and **Figure 5-4** could be used by the pattern shop, foundry and machine shop. Notice that larger tolerances are applied to the rough-to-rough and finish-to-rough dimensions and that a minimum number of finish-to-rough dimensions is used. In addition, the general notes include finish, as well as the draft, casting radii and material specifications found on the casting and machining drawings.

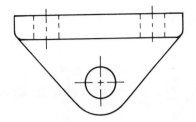

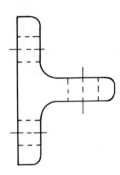
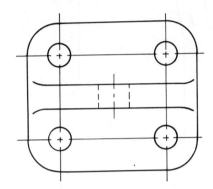

PROBLEM 5-3. Casting/machining drawing. Remove and tape down the duplicate worksheet at the back of the book. Add finish marks and dimension the part completely, using a minimum of finish-to-rough dimensions. Apply a ± .04 tolerance to finish-to-rough and rough-to-rough dimensions. Specify a ± .02 tolerance for finish-to-finish dimensions, fillets, rounds, 2° draft and .12 FINISH in a general note.

5-3 LEARNING APPLICATION

The following problems should show how well you have mastered the module objectives.

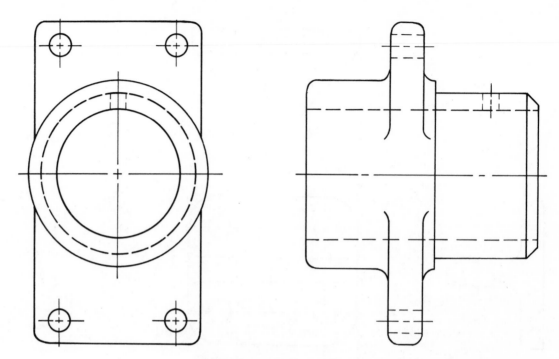

PROBLEM 5-4. Shaft support casting (part number 134). Tape down a drawing sheet. Draw and dimension in millimeters the casting needed to produce the given machined part. Add an extra 3mm of material to all machined surfaces. (For diameters, this means adding 3mm on each side so the cast **OD** would be 6mm larger than the machined OD, while the cast **ID** would be 6mm smaller). Specify 2° draft, ± 1mm tolerances, and ASTM A4864 Gray Iron material.

PROBLEM 5-5. Shaft support (part number 135). Tape down a drawing sheet. Draw and dimension the machining drawing version for the part shown in Problem 5-4. Include finish marks and specify ± 1mm tolerances for finish-to-rough dimensions and ± 0.5 for finish-to-finish.

PROBLEM 5-6. Intake manifold. Tape down a drawing sheet. Make a casting/machining drawing, dimensioned in decimal inches. Use the tolerances, draft and finish specified for Problem 5-3. In order to clearly describe the part, it is suggested that you add a side view and a sectional view cut through the axis of symmetry. The sectional view would clarify the central portion and make it possible to minimize the number of hidden lines needed in the side view.

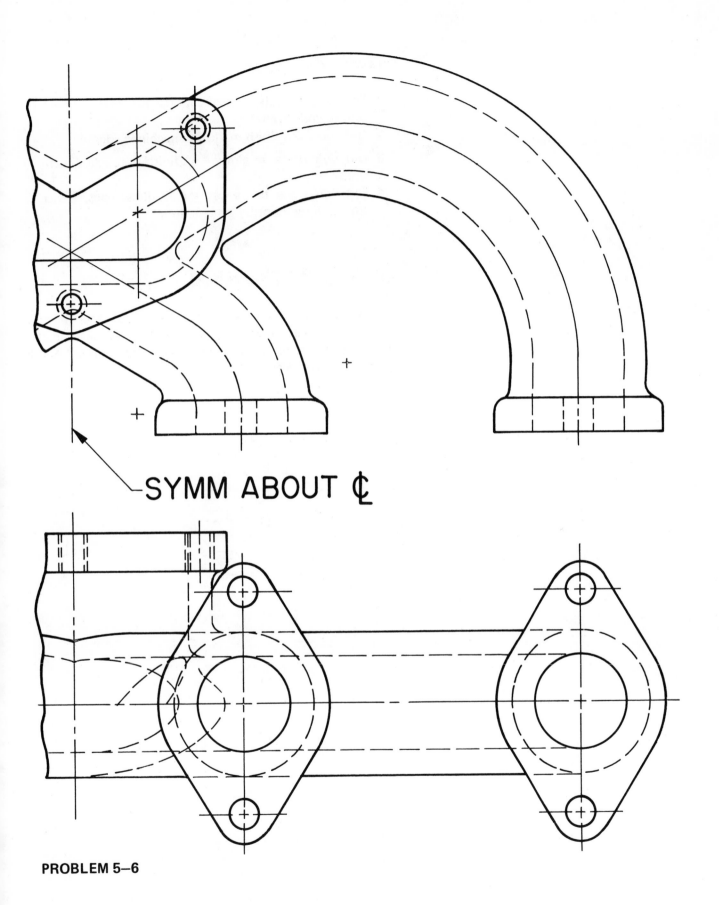

SYMM ABOUT ℄

PROBLEM 5-6

DRAWING CHECKLIST

1 Did you include specifications for draft, fillets, rounds and finish?

2. Did you show finish marks on machined edges?

3. Did you use a minimum of finish-to-rough dimensions?

4. Did you show tolerances for all dimensions, either applied directly or by general note?

ANSWERS TO PROBLEMS

These partial solutions contain some of the general dimensions and a general approach which may be used. In some cases, other approaches may also satisfy ANSI dimensioning requirements.

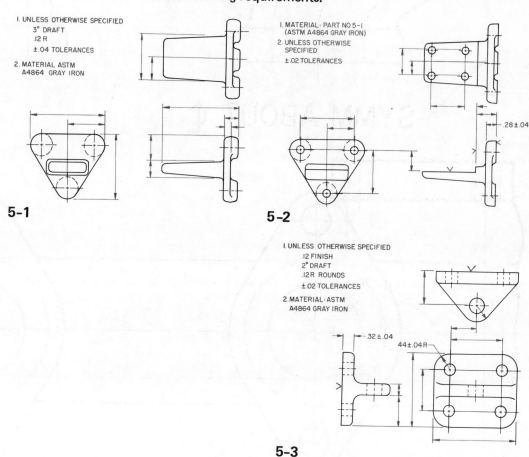

1. UNLESS OTHERWISE SPECIFIED
   3° DRAFT
   .12 R
   ±.04 TOLERANCES

2. MATERIAL ASTM
   A4864 GRAY IRON

5-1

1. MATERIAL: PART NO 5-1
   (ASTM A4864 GRAY IRON)

2. UNLESS OTHERWISE
   SPECIFIED
   ±.02 TOLERANCES

.28 ±.04

5-2

1. UNLESS OTHERWISE SPECIFIED
   .12 FINISH
   2° DRAFT
   .12R ROUNDS
   ±.02 TOLERANCES

2. MATERIAL: ASTM
   A4864 GRAY IRON

.32 ±.04

44±.04R

5-3

6

SHEET METAL

The dimensioning of a part must permit the parts to be assembled and to work as intended, but this dimensioning should also permit the economical manufacture of the parts. These requirements present a special challenge in the dimensioning of sheet metal parts because the manufacturing processes used make it necessary to apply bigger than normal tolerances to dimensions going to bends and across bends. This means that the location of holes and other critical features is less accurate with sheet-metal forming than with other processes, and much greater care is needed to prevent accumulation of tolerances and rejection of the produced part.

PURPOSE

The purpose of this module is to show you how to dimension sheet metal parts so that the requirements of function and assembly can be met while using conventional production methods.

PREREQUISITES

Before you begin this module you should be able to:

1. Use drafting instruments and apply multiview drawing principles (Worktexts 1 and 2).
2. Use proper techniques to dimension holes, radii and angles (Modules 2 and 3).
3. Apply fundamental dimensioning rules (Modules 1 and 4).

OBJECTIVES

Upon completion of this module, you will be able to:

1. Draw and dimension a formed sheet metal part, observing the rules given in ANSI Y14.5, paragraph 1.7a (Tolerancing), paragraph 1.7c (Relationships), paragraph 1.7d (Tolerance Accumulation) and paragraph 1.7g (Profile View).
2. Calculate the blank length of a formed sheet metal part.
3. Determine a formed shape when given the blank shape and the bending dimensions.
4. Determine the blank shape when given the formed shape.

EQUIPMENT AND REFERENCES

Before you begin this module, you should have:

1. All normal drafting equipment, including circle template and dividers.
2. Calculator (optional).
3. ANSI Y14.5, *Dimensioning and Tolerancing*.
4. ANSI Y14.10, *Sheet Metal Stampings* (optional reference).

6-1 MANUFACTURING METHODS

When drawing and dimensioning a sheet metal part, it is important to consider the manufacturing processes. The selection of the process depends on the type of part and the number of parts to be produced.

CUTTING PROCESSES

Shearing is often the least expensive way of cutting sheet metal, partially because there is no scrap with this process (**Figure 6-1**). If the width of the part is the same as an available stock width, a *parting-type cutoff* may be used. If the part is a high production item, a *blanking die* may be the best choice.

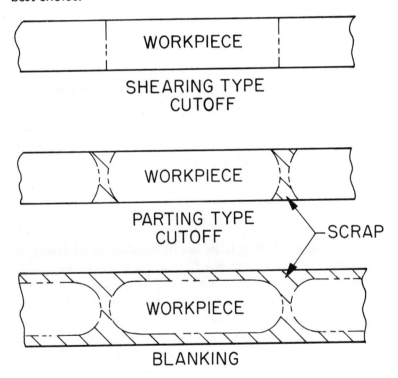

Figure 6-1. Diagram showing how the three major cutting processes produce different workpieces.

BENDING PROCESSES

The method to be used for bending a part depends on the quantity to be produced and the complexity of the part (**Figure 6-2**). For low production quantities, a general purpose bending die or a sheet metal brake may be used. For high production, a special forming die can be built which will perform all of the bends at once and rapidly produce large quantities of the same sheet metal part.

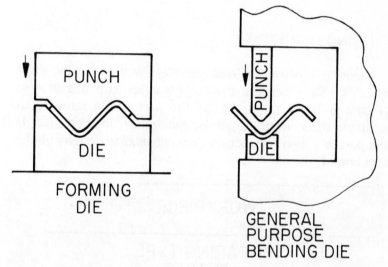

Figure 6-2. Illustration of two major bending processes used for forming sheet metal.

6-2 FORMED PARTS

HOLE DIMENSIONS

Regardless of whether holes are to be punched or drilled, it is normally best for both the function and the manufacture of the part if the holes are tied together (located from each other or from a common reference point) as in **Figure 6-3**. The most important relationships are usually between the holes and these dimensions are the controlling guidelines for the design of either punching dies or drill jigs.

Since bending is a relatively inaccurate process that is performed after the holes are punched, a more liberal tolerance is necessary on dimensions between holes that are separated by bends. This tolerance will vary from company to company.

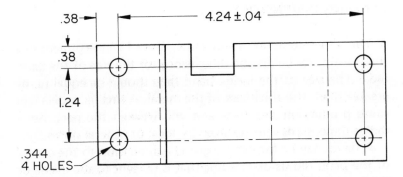

Figure 6-3. Hole dimensions. The holes are "tied together" with the 1.24 and 4.24 dimensions with a more liberal tolerance on the latter because the holes are separated by bends. The missing dimensions for this part are shown in **Figure 6-4** and **Figure 6-5**.

BLANKING AND SHEARING DIMENSIONS

Since cutting normally takes place before bending, it is helpful for manufacturing if all the cut features are tied together. Tolerance accumulation can be minimized if all the features are located from the same end of the part (**Figure 6-4**). The dimensions that are affected by bends require a more liberal tolerance. Dimensions across bends should be kept to a minimum to avoid accumulation of tolerances.

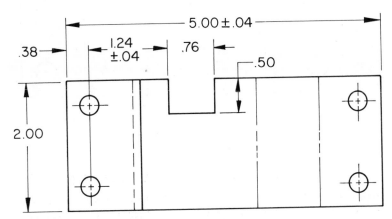

Figure 6-4. Blanking and shearing dimensions. The cutout features are "tied together" by giving the size of the .76 X .50 cutout directly rather than by using two separate dimensions across the bend. This simplifies the designing of the blanking die. The 1.24 dimension requires a more liberal tolerance because it goes across a bend.

BENDING DIMENSIONS

The dimensions for bending operations include dimensions that locate the bends and dimensions for the size of the bend radii. The size of the inside bend radii should be equal to, or greater than, the thickness of the metal. A smaller radius will cause production problems and will weaken the part. Bend radius dimensions may be given by local or general notes.

A bend may be located by giving a dimension to the center of the bend radius or to a line that is tangent to the inside or the outside bend radius (**Figure 6-5**). The tangency method is best when locating a horizontal or vertical surface. The advantage of locating by dimensioning to the center of a bend radius is that the same location dimension applies to both the top and bottom half of the forming die. Regardless of the method of locating the bends, the dimensions must be given in the profile view.

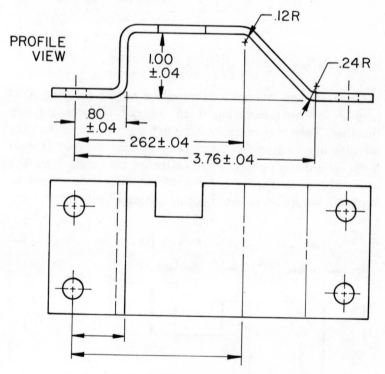

WRONG VIEW FOR LOCATING BENDS

Figure 6-5. Bending dimensions. This drawing shows the bending dimensions for the part previously shown in **Figure 6-3** and **Figure 6-4**. The complete drawing would show the hole dimensions, blanking dimensions and bending dimensions. The bends should be located in the front view rather than the bottom view because in this case, the profiles of all the bends are shown in the front view.

Since bending is performed independently of cutting, bend locations should be tied together directly or through a common baseline as shown in **Figure 6-5**. Another help to manufacturing is to give all the dimensions to the same side of the sheet metal so that these same dimensions may apply to the same side of the forming die. Always dimension to the most important side of the metal (usually the side that touches the mating part).

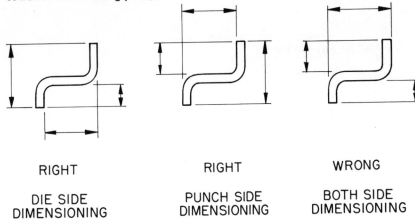

RIGHT	RIGHT	WRONG
DIE SIDE DIMENSIONING	PUNCH SIDE DIMENSIONING	BOTH SIDE DIMENSIONING

Figure 6-6. Dimensioning side. Dimensions should go to either the punch side or the die side of the metal; not to both sides.

PROBLEM 6-1. Dimensioning formed sheet metal parts. Remove and tape down the duplicate worksheets at the back of the book. Dimension each part completely, using ±.04 tolerance for dimensions to or across bends.

PROBLEM 6–1A

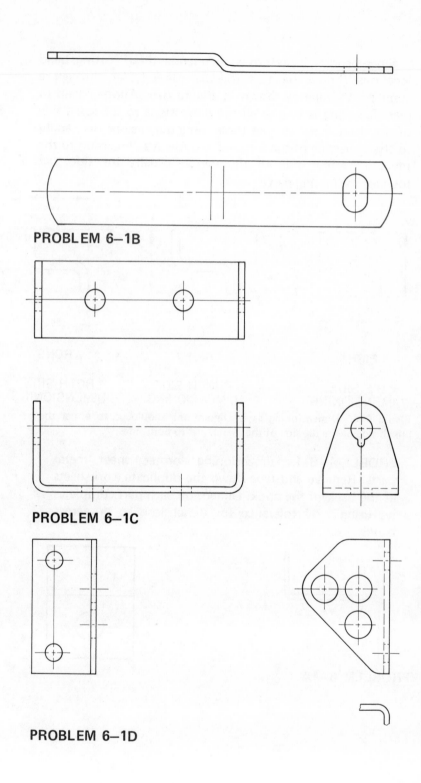

PROBLEM 6–1B

PROBLEM 6–1C

PROBLEM 6–1D

6-3 SHEET METAL BLANKS

CALCULATING BLANK LENGTHS

Many companies require that a sheet metal drawing show the length of the part before bending. The steps involved in finding this blank length are (1) finding the straight lengths, (2) calculating the bend allowance and (3) adding the straight lengths to the bend allowances for the total.

Straight Lengths. Straight lengths refer to the distance before, between, and after the bends in a part. Since the dimensioning may not be between centers for the bend radii, it may be necessary to apply some mathematics or carefully scale the drawing to find the distance between centers and accurately determine the straight lengths so the sizes correspond with the given dimensions.

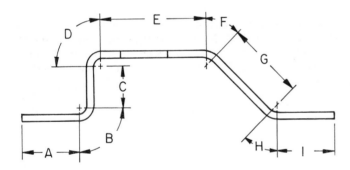

Figure 6-7. Straight lengths and bends. To find the blank length of a part like that shown in previous examples, the first step must be to find the straight lengths between the bends. This requires finding the exact centers of all the bend radii and accurately drawing extension lines from these points and perpendicular to the surfaces, through the points of tangency between the straight lines and arcs. In this illustration, **A, C, E, G** and **I** represent the straight lengths while **B, D, F** and **H** represent the bends.

Bend Allowances. Bend allowances refer to the amount of material needed to make the bends. There are a number of formulas that are used to find the bend allowance and they all give similar results. The *Machinery's Handbook* contains formulas and tables for finding the bend allowances for right-angle bends for several types of materials with different thicknesses and inside bend radii. Since these apply at 90°, ratios must be used to find bend allowances for larger and smaller bends. For example, a bend through 30° takes 1/3 of the material of a 90° bend.

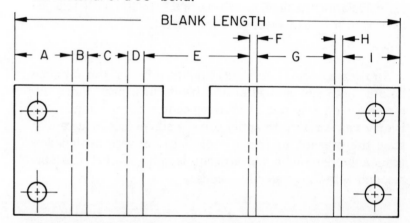

Figure 6-8. Blank length. The amount of material required for 90° bends like **B** and **D** is found by applying a formula such as **BA = (.64 X material thickness) + (1.57 X inside bend radius)**. Since **F** and **H** are 45° bends, they would require half as much material as 90° bends. The blank length is found by adding all straight lengths and bend allowances.

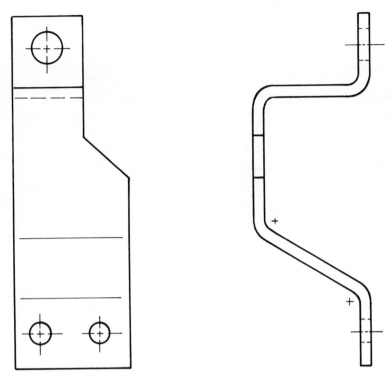

PROBLEM 6-2. Calculating blank lengths. Find the blank length of the object, using the formula Bend Allowance = (.64 X Material Thickness) + (1.57 X Inside Bend Radius).

BLANK VIEW

While the usual procedure is to show the size and shape of the formed part, in some cases it is more practical to show the blank. If the bends are not perpendicular to a cut edge, a complicated blank is necessary to produce a simple looking formed part. In many cases, a very simple blank can be drawn, and although the formed part will look a little different, the functional requirements will still be met. This is desirable because a simpler blank means simpler tooling and lower cost. See **Figure 6-9** and **Figure 6-10**.

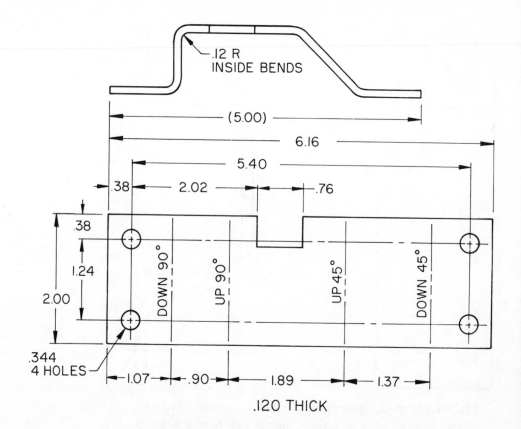

Figure 6-9. Dimensioned blank view. This drawing shows the blank view approach to dimensioning the part that was dimensioned in formed views in **Figures 6-3, 6-4,** and **6-5.** This method is the most practical for simple parts that are made on general purpose equipment rather than special dies. Cut features between bends are dimensioned from the end but are originally located with respect to the original bend lines as shown in **Figure 6-8.** The dimensioned bend lines are halfway between the lines at the start and end of the bends as shown in **Figure 6-8.**

PROBLEM 6-3. Dimensioning a blank view. Tape down a drawing sheet. Draw and dimension the blank that would give the end result shown in Problem 6-2.

Figure 6-10. Partial blank view. Although parts **A** and **B** are very similar and would perform the same function, **A** would be preferable because it would be easier to produce. All sides of the blank for **A** are straight lines, allowing the use of either shears or a simple die to produce the blank. Since one side of **B** is irregular, shearing would not be possible and a complex blanking die would be required.

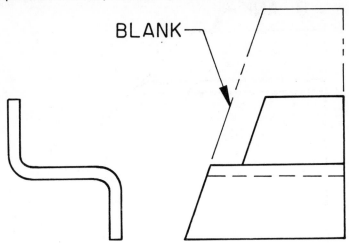

BLANK

SIMPLE TO PRODUCE

(A)

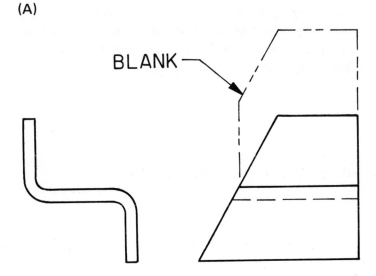

BLANK

MORE DIFFICULT

(B)

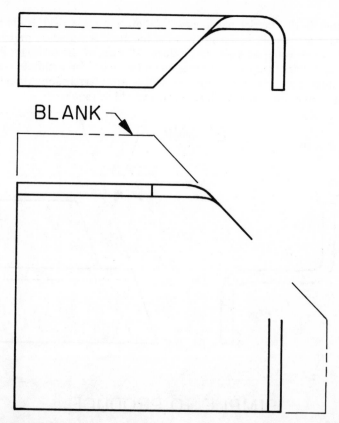

BLANK

PROBLEM 6-4. Partial blank view. Remove and tape down the duplicate worksheet at the back of the book. Complete the front view and add the side view according to the finished portions.

6-4 LEARNING APPLICATION

The drawings in Problem 6-5 will show how well you have met module objectives 1 and 2.

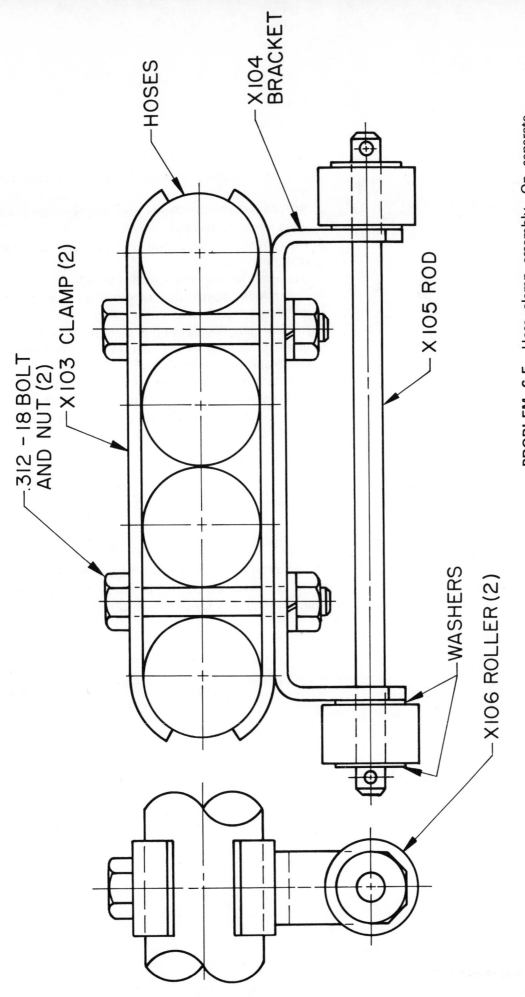

HOSES

X104 BRACKET

.312 – 18 BOLT AND NUT (2)

X103 CLAMP (2)

X105 ROD

WASHERS

X106 ROLLER (2)

PROBLEM 6-5. Hose clamp assembly. On separate drawing sheets, draw and dimension the clamp and the bracket, scaling the layout for sizes.

DRAWING CHECKLIST

1. Did you dimension between holes?
2. Are bends dimensioned to the important side of the metal?
3. Are dimensions to bends and across bends given a more liberal tolerance?
4. Are parts dimensioned to provide for mating features?
5. Are bends located in the profile view?
6. Are bends located by tangency to horizontal or vertical surfaces?
7. In blank length calculations, were the straight lengths to the tangencies of the bend radii rather than to the intersections of the surfaces?
8. Are the bend angles used for bend allowance calculations based on the degrees through which the metal is bent rather than the degrees between the intersecting surfaces?

ANSWERS TO PROBLEMS

These partial solutions contain some of the dimensions and a general approach which may be used. In some cases, other approaches may also satisfy the ANSI dimensioning requirements.

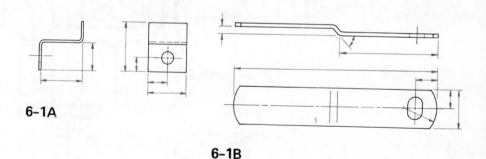

6-1A

6-1B

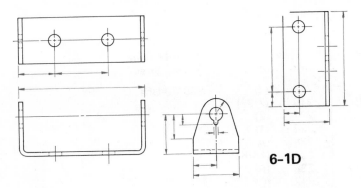

6–1C

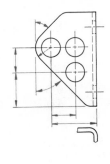

6–1D

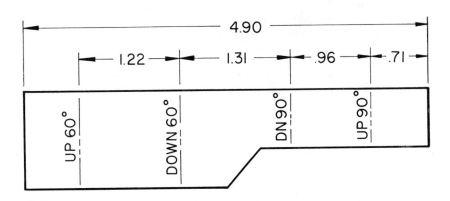

6–2

Inch Drill Sizes		Unified National Thread Sizes	Inch Drill Sizes		Unified National Thread Sizes
.078	(5/64)		.213	(3)	.250–28 UNF
.078	(47)	.099 (3)–47 UNC	.219	(7/32)	
.081	(46)		.221	(2)	
.082	(45)	.099 (3)–56 UNF	.228	(1)	
.086	(44)		.234	(A)	
.089	(43)	.112 (4)–40 UNC	.234	(15/64)	
.094	(42)	.112 (4)–42 UNF	.238	(B)	
.094	(3/32)		.242	(C)	
.096	(41)		.246	(D)	
.098	(40)		.250	(1/4)	
.100	(39)		.250	(E)	
.102	(38)	.125 (5)–40 UNC	.257	(F)	.312–18 UNC
.104	(37)	.125 (5)–44 UNF	.261	(G)	
.106	(36)	.138 (6)–32 UNC	.266	(17/64)	
.109	(7/64)		.266	(H)	
.110	(35)		.272	(I)	.312-24 UNF
.111	(34)		.277	(J)	
.113	(33)	.138 (6)–40 UNF	.281	(K)	
.116	(32)		.281	(9/32)	
.120	(31)		.290	(L)	
.125	(1/8)		.295	(M)	
.128	(30)		.297	(19/64)	
.136	(29)	.164 (8)–32 UNC	.302	(N)	
.140	(9/64)	.164 (8)–36 UNF	.312	(5/16)	.375–16 UNC
.141	(28)		.316	(O)	
.144	(27)		.323	(P)	
.147	(26)		.328	(21/64)	
.150	(25)	.190 (10)–24 UNC	.332	(Q)	.375–24 UNF
.152	(24)		.339	(R)	
.154	(23)		.344	(11/32)	
.156	(5/32)		.348	(S)	
.157	(22)		.358	(T)	
.159	(21)	.190 (10)–32 UNF	.359	(23/64)	
.161	(20)		.368	(U)	.438–14 UNC
.166	(19)		.375	(3/8)	
.170	(18)		.377	(V)	
.172	(11/64)		.386	(W)	
.173	(17)		.391	(25/64)	.438-20 UNF
.177	(16)	.216 (12)–24 UNC	.397	(X)	
.180	(15)		.404	(Y)	
.182	(14)	.216 (12)–28 UNF	.406	(13/32)	
.185	(13)		.413	(Z)	
.188	(3/16)		.422	(27/64)	.500–13 UNC
.189	(12)		.438	(7/16)	
.191	(11)		.453	(29/64)	.500–20 UNF
.194	(10)		.469	(15/32)	
.196	(9)		.484	(31/64)	.562–12 UNC
.199	(8)		.500	(1/2)	
.201	(7)	.250–20 UNC	.516	(33/64)	.562–18 UNF
.203	(13/64)		.531	(17/32)	.625–11 UNC
.204	(6)		.547	(35/64)	
.206	(5)		.562	(9/16)	
.209	(4)		.578	(37/64)	.625–18 UNF

Inch Drill Sizes		Unified National Thread Sizes	Inch Drill Sizes		Unified National Thread Sizes
.594	(19/32)		.812	(13/16)	.875–14 UNF
.609	(39/64)		.828	(53/64)	
.625	(5/8)		.844	(27/32)	
.641	(41/64)		.859	(55/64)	
.656	(21/32)	.750–10 UNC	.875	(7/8)	1.000–12 UNC
.672	(43/64)		.891	(57/64)	
.688	(11/16)	.750–16 UNF	.906	(29/32)	
.703	(45/64)		.922	(59/64)	1.000–12 UNF
.719	(23/32)		.938	(15/16)	
.734	(47/64)		.953	(61/64)	
.750	(3/4)		.969	(31/32)	
.766	(49/64)	.875–9 UNC	.984	(63/64)	1.125–7 UNC
.781	(25/32)		1.000	(1)	
.797	(51/64)				

APPENDIX II

Millimeter Drill Sizes	ISO Metric Threads*	Millimeter Drill Sizes	ISO Metric Threads
3.00		11.80	
3.20	M4 X 0.7	12.00	
3.50		12.20	
4.00		12.50	
4.20	M5 X 0.8	12.80	
4.50		13.00	
4.80		13.20	
5.00	M6 X 1	13.50	
5.20		13.80	
5.50		14.00	M16 X 2
5.80		14.25	
6.00		14.50	
6.20		14.75	
6.50		15.00	
6.80	M8 X 1.25	15.25	
7.00		15.50	
7.20		15.75	
7.50		16.00	
7.80		16.25	
8.00		16.50	
8.20		16.75	
8.50	M10 X 1.5	17.00	
8.80		17.25	
9.00		17.50	M20 X 2.5
9.20		17.75	
9.50		18.00	
9.80		18.25	
10.00		18.50	
10.20	M12 X 1.75	18.75	
10.50		19.00	
10.80		19.25	
11.00		19.50	
11.20		19.75	
11.50		20.00	

*International Standards Organization Metric Threads.

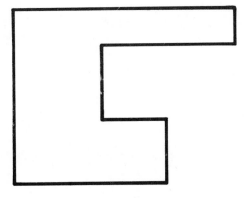

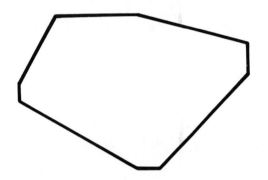

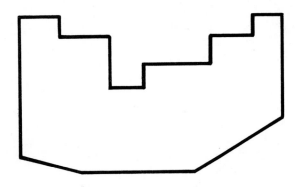

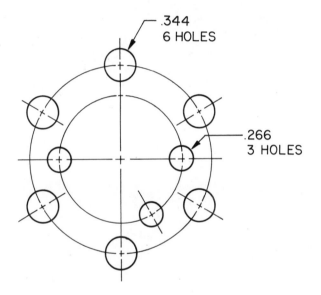

.344
6 HOLES

.266
3 HOLES

PROBLEM 2—2

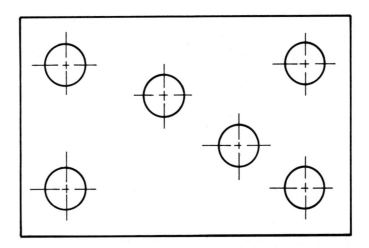

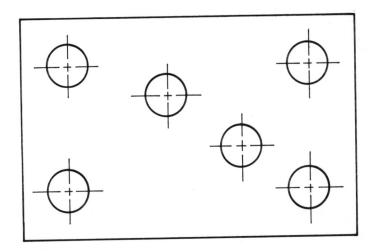

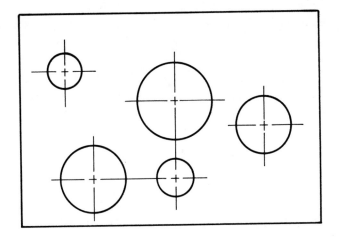

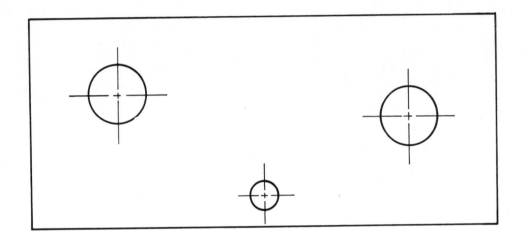

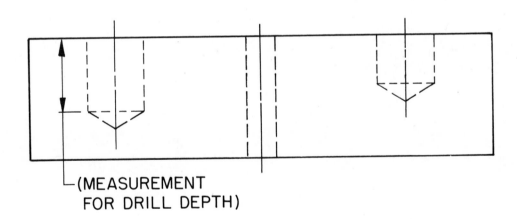

⌐(MEASUREMENT
FOR DRILL DEPTH)

PROBLEM 2—7

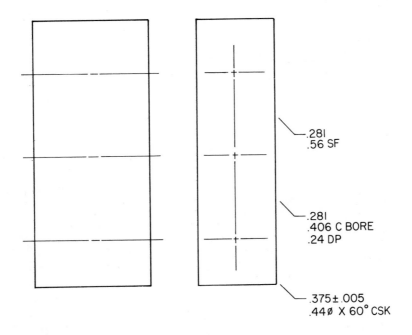

.281
.56 SF

.281
.406 C BORE
.24 DP

.375±.005
.44Ø X 60° CSK

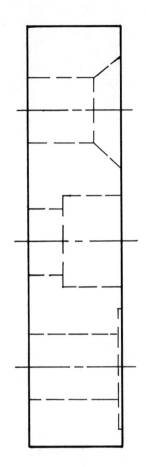
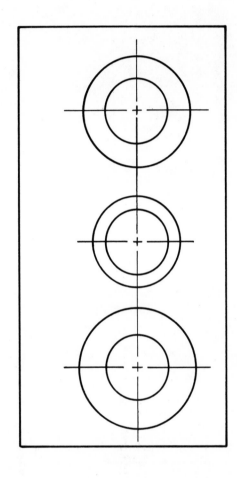

PROBLEM 2—9

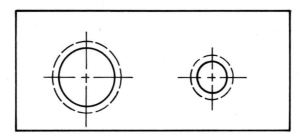

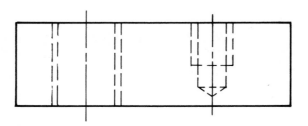

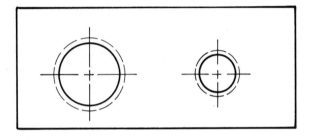

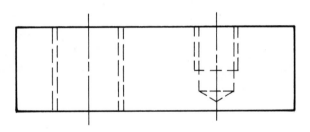

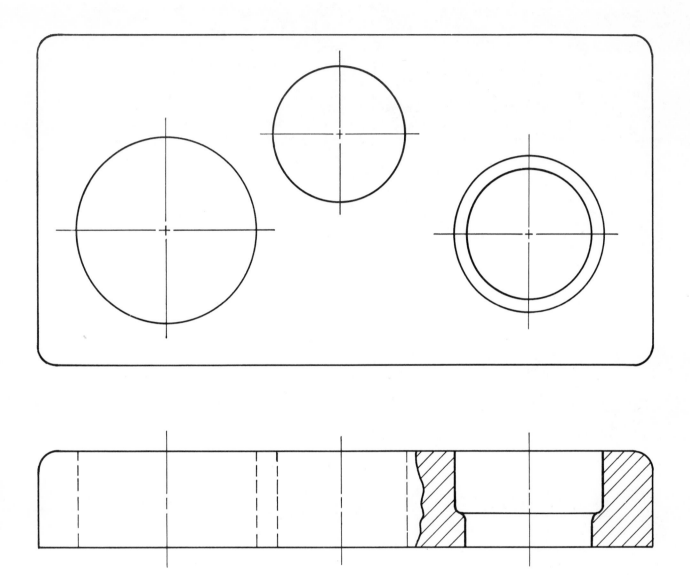

PROBLEM 3–1

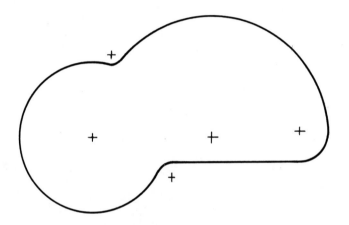

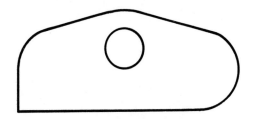

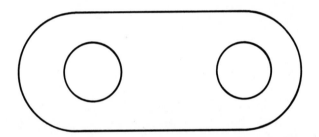

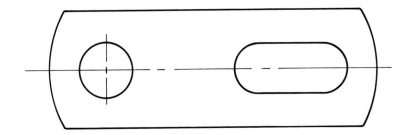

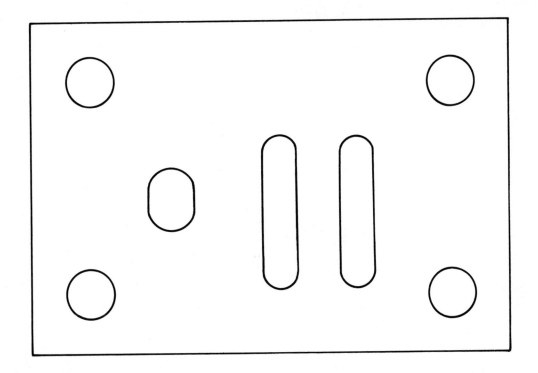

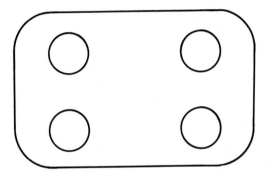

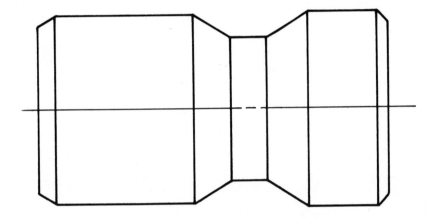

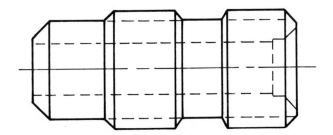

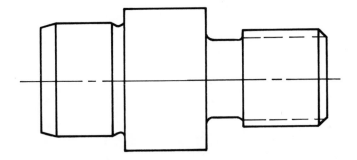

PROBLEM 3–8A

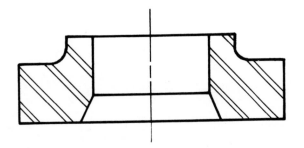

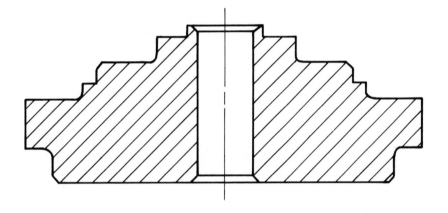

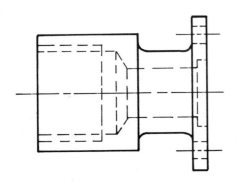

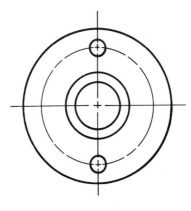

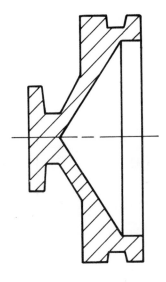

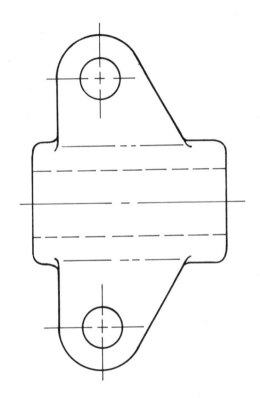
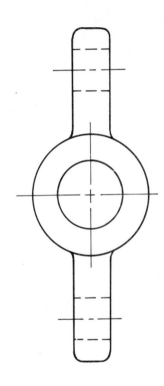

PROBLEM 4—2

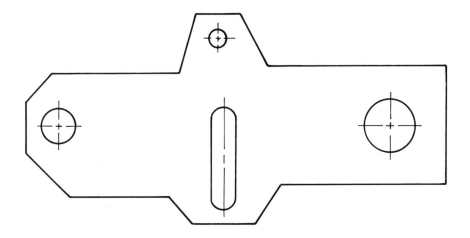

PROBLEM 4—3A

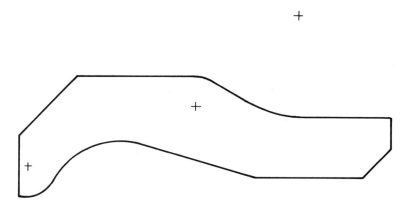

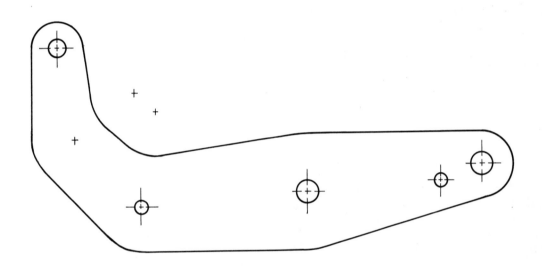

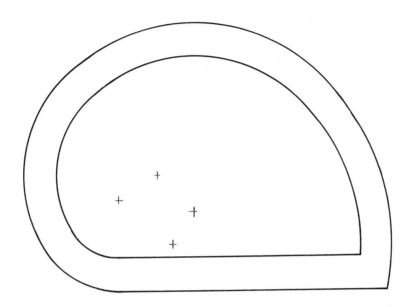

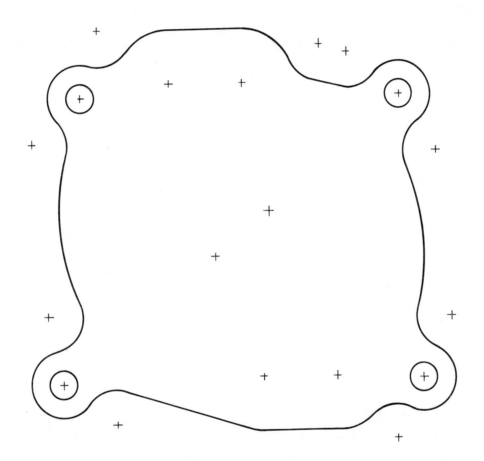

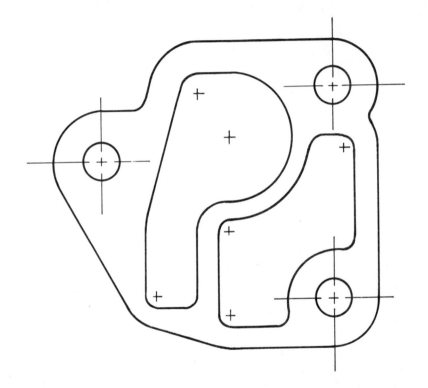

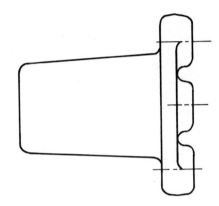

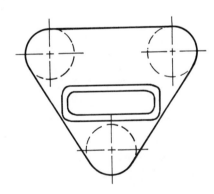

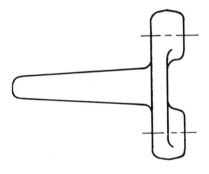

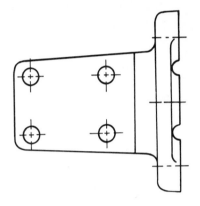

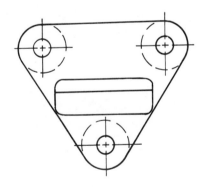

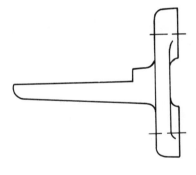

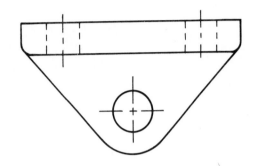

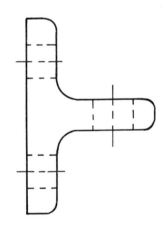

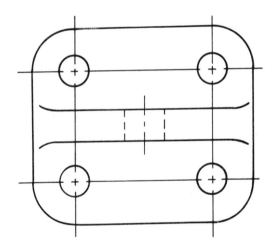

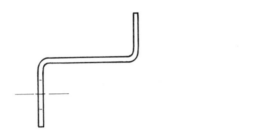
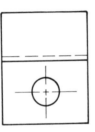

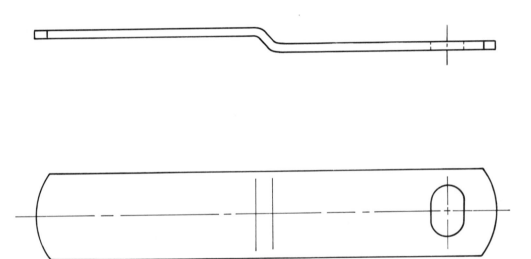

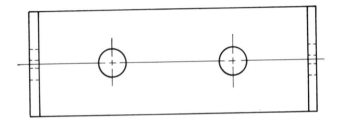

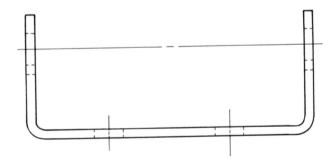

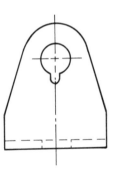

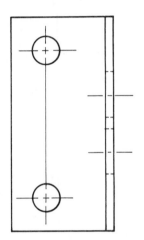

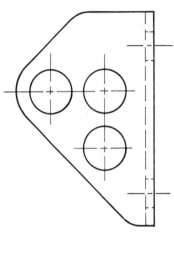

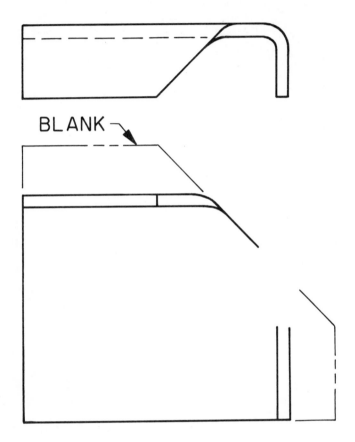

BLANK